W9-BZJ-970

Universitext

Universitext

Editors (North America): J.H. Ewing, F.W. Gehring, and P.R. Halmos

(continued after index)

George A. Jennings

Modern Geometry with Applications

With 150 figures

Springer-Verlag

New York Berlin Heidelberg London Paris
Tokyo Hong Kong Barcelona Budapest

George A. Jennings
Department of Mathematics
California State University
Dominguez Hills
1000 E. Victoria Blvd.
Carson, CA 90747 USA

Mathematics Subject Classifications (1991): 51M04, 51M09

Library of Congress Cataloging-in-Publication Data
Jennings, George, 1951 -
    Modern geometry with applications / George Jennings.
        p.   cm. -- (Universitext)
    Includes bibliographical references and index.
    ISBN 0-387-94222-X
    1. Geometry, Modern.   I. Title.   II.Series.
    QA473.J46   1994
    516 .02--dc20   93-278181

Printed on acid-free paper.

Production managed by Karen Phillips, manufacturing supervised by Jacqui Ashri.
Camera-ready copy prepared from the author's TEX files.
Printed and bound by Edwards Brothers, Inc., Ann Arbor, MI.
Printed in the United States of America.

9 8 7 6 5 4 3 2 1

ISBN 0-387-94222-X Springer-Verlag New York Berlin Heidelberg
ISBN 3-540-94222-X Springer-Verlag Berlin Heidelberg New York

Foreword

This book is an introduction to the theory and applications of "modern geometry" – roughly speaking, geometry that was developed after Euclid. It covers three major areas of non-Euclidean geometry and their applications: spherical geometry (used in navigation and astronomy), projective geometry (used in art), and spacetime geometry (used in the Special Theory of Relativity). In addition it treats some of the more useful topics from Euclidean geometry, focusing on the use of Euclidean motions, and includes a chapter on conics and the orbits of planets.

My aim in writing this book was to balance theory with applications. It seems to me that students of geometry, especially prospective mathematics teachers, need to be aware of how geometry is used as well as how it is derived. Every topic in the book is motivated by an application and many additional applications are given in the exercises. This emphasis on applications is responsible for a somewhat nontraditional choice of topics: I left out hyperbolic geometry, a traditional topic with practically no applications that are intelligible to undergraduates, and replaced it with the spacetime geometry of Special Relativity, a thoroughly non-Euclidean geometry with striking implications for our own physical universe. The book contains enough material for a one semester course in geometry at the sophomore-to-senior level, as well as many exercises, mostly of a non-routine nature (the instructor may want to supplement them with routine exercises of his/her own).

I prepared the illustrations on a PC using *Windows Draw 3.0* by Micrografx and *Mathematica 2.2* by Wolfram Research.

Contents

1
Euclidean Geometry

1.1 Euclidean Space

Euclidean space is the space that contains the ordinary objects of high school geometry: lines, circles, spheres, and so on. An n-dimensional Euclidean space is essentially the same thing as $\mathbf{R}^n$, the set of all ordered n-tuples $(x_1, \ldots, x_n)$ of real numbers ($\mathbf{R}$ stands for 'Real numbers', n for 'n-dimensonal'). The notation $\mathbf{E}^n$ stands for n-dimensional Euclidean space ($\mathbf{E}$ for Euclid, n for 'n-dimensional'.) $\mathbf{R}^n$ and $\mathbf{E}^n$ differ in that $\mathbf{R}^n$ comes equipped with a special system of coordinates and a specially marked point (the origin), while Euclidean space has no natural coordinates or distinguished points. In this chapter we shall refer to geometrical space as $\mathbf{R}^n$ when we are using coordinates and $\mathbf{E}^n$ when we are not.

Euclidean space is absolutely uniform (the technical term is "homogeneous")—every place in a Euclidean space looks the same as every other place. This uniformity of Euclidean space is a key feature of Euclidean geometry, for it enables one to move objects around inside Euclidean space without bending, stretching, or otherwise distorting them. Distant objects can be compared without changing their shapes by bringing them together and placing one on top of the other. Reliable measurement is possible in Euclidean space because one can move measuring instruments from place to place without destroying their accuracy. The most important measuring tools for the Euclidean geometer are a *ruler* (for measuring distances), the *protractor* (for measuring angles), and a sense of *orientation* or rotational direction for distinguishing between clockwise and counterclockwise rotations.

Objects are moved from place to place in space by the action of *functions* (Fig. 1.1). A function

$$f : \mathbf{E}^n \to \mathbf{E}^n$$

(read "f is a function from $\mathbf{E}^n$ to $\mathbf{E}^n$") takes each point $P \in \mathbf{E}^n$ and moves it to a new position $f(P) \in \mathbf{E}^n$. [1] If S is a set of points then

$$f(S) = \{f(P) \mid P \in S\}$$

is the set that results from applying the function f to all of the points in S.

1.2   Isometries and Congruence

The notation PQ means "the distance from the point P to the point Q". The most important functions on $\mathbf{E}^n$ are those that preserve distance.

Definition 1.2.1 Isometry. A function $f : \mathbf{E}^n \to \mathbf{E}^n$ is an *isometry* if, for all points P and Q in $\mathbf{E}^n$,

$$f(P)f(Q) = PQ.$$

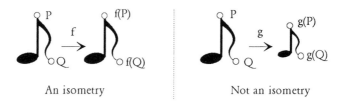

An isometry   Not an isometry

FIGURE 1.1. Functions on Euclidean space.

Isometries on Euclidean space are often called "Euclidean isometries", "Euclidean motions", or "Euclidean transformations".

All the familiar geometric qualities of figures: length, area, volume, the size of angles, etc., are derived from distances. The length of a polygon is the sum of the distances between its adjoining vertices and the lengths of more general curves are computed by approximating them with polygons. Angles are measured (in radians) by measuring the length of arc subtended by the angle on a unit circle. The area of a rectangle is the product of its length and its width, the volume of a rectangular box is (length)×(width)×(height), and the areas and volumes of more general regions can be approximated by filling them up with little rectangles or boxes.

---

[1] It may happen that $f(P) = P$.

Since isometries preserve distances they also preserve lengths, angles, areas, and volumes. In short, *isometries preserve the size and shape of every geometric figure.*

Example. The line segment $\overline{PQ}$ is the shortest curve connecting P to Q. If f is an isometry then it follows that $f(\overline{PQ})$ must be the shortest curve connecting f(P) to f(Q), since isometries preserve lengths. Therefore $f(\overline{PQ})$ also is a line segment

$$f(\overline{PQ}) = \overline{f(P)f(Q)}.$$

Definition 1.2.2 Congruence. Subsets $A, B \subseteq \mathbf{E}^n$ are *congruent* (in symbols $A \cong B$) if there is an isometry f such that $f(A) = B$.

Exercise 1.2.1 a). Show that every isometry is a one-to-one function. (In other words, show that if f is an isometry and P and Q are points with $P \neq Q$ then $f(P) \neq f(Q)$).

b) Assume that f is an isometry and that it has an inverse function f^{-1}. Show that f^{-1} also is an isometry. (Exercise 1.10.2 shows that every isometry does, in fact, have an inverse).

Exercise 1.2.2 Let $C \subset \mathbf{E}^2$ be the circle with center P and radius r. Prove that if $f : \mathbf{E}^2 \to \mathbf{E}^2$ is an isometry then $f(C)$ is the circle with center $f(P)$ and radius r. (Hint: prove that if C' is the circle with center $f(P)$ and radius r then $f(C) \subseteq C'$ and $f^{-1}(C') \subseteq C$. You may assume that f has an inverse and that f^{-1} is an isometry).

1.3 Reflections in the Plane

Every Euclidean isometry is a combination of three fundamental types: reflections, translations, and rotations.

Let P be a point and L a line in $\mathbf{E}^2$. Drop a perpendicular M from P to L. The *reflection* of P in L is the point $P' \in M$ such that P' lies on the opposite side of L from P and P' is the same distance from L as P. ($P' = P$ if $P \in L$; see Fig. 1.2).

To reflect an entire figure simply reflect all of its points.

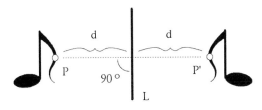

FIGURE 1.2. Reflection in a Line.

There are three other ways to reflect a figure, all leading to the same result.

1. Flip the plane over onto its back by rotating it $180°$ around L in three dimensional space. Equivalently, fold the plane over along the line L, then trace the figure on the other side. (See Fig. 1.3. This construction explains why reflections are isometries, since it is obvious that merely rotating a plane does not change the size or shape of any figure in it).

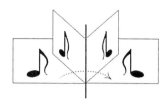

FIGURE 1.3. Reflecting by flipping over.

2. A "ruler and compass construction". If $P \notin L$, draw two circles with their centers on L, intersecting at P (any two such circles will do). The two circles will intersect at another point on the opposite side of L; this point is the reflection of P (Fig. 1.4.).

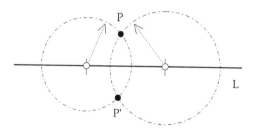

FIGURE 1.4. Ruler and compass construction.

The ruler and compass construction is based on the fact that reflection in a line L preserves any circle that is centered on L. Let C be a circle with center $Z \in L$ and let f be the reflection in L. $f(Z) = Z$, since $Z \in L$, so $f(Q)Z = QZ$ for every point Q, since f preserves distances. In particular if $Q \in C$ then $f(Q)$ must also lie on C. This holds for both circles in the ruler and compass construction, so f maps each point in their intersection to another point in their intersection.

3. Reflection in the x axis in $\mathbf{R}^2$ is given by the formula

$$f(x, y) = (x, -y).$$

Exercise 1.3.1 Find a formula in terms of $a, b,$ and c for reflection in an arbitrary line $aX + bY + c = 0$ in $\mathbf{R}^2$.

Exercise 1.3.2 a) Prove that any two circles in the plane intersect in exactly two, one, or zero points depending on whether the distance between their centers is less than, equal to, or greater than the sum of their radii.

b) Prove that if two circles in $\mathbf{E}^2$ intersect in two points then the line connecting their centers is the perpendicular bisector of the line segment connecting the points where the two circles intersect.

(Hint. Write equations for the circles and solve them simultaneously. This is not hard to do if you set up coordinates so that the x axis passes through the centers of both circles.)

1.4 Reflections in Space

Let P be a point and H a plane in $\mathbf{E}^3$. Drop a perpendicular M from P to H. The reflection of P in H is the point $P' \in M$ such that P' lies on the opposite side of H from P and P' is the same distance from H as P ($P' = P$ if $P \in H$; see Fig. 1.5).

One way to find P' would be to construct three spheres with centers on H, all intersecting at P. In $P \notin H$ then the spheres intersect at exactly two points, P and P' (Why?).

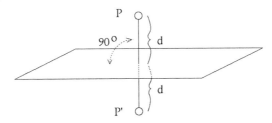

FIGURE 1.5. Reflection in a plane.

Reflection in the x, y plane in $\mathbf{R}^3$ is given by the formula

$$f(x, y, z) = (x, y, -z).$$

Exercise 1.4.1 Show that every reflection (in $\mathbf{E}^2$ or $\mathbf{E}^3$) is its own inverse: if $f : \mathbf{E}^n \rightarrow \mathbf{E}^n$ is a reflection (n = 2 or 3) then $f^{-1} = f$.

Vectors

A *vector* is a directed line segment that points from one point to another in $\mathbf{E}^n$ or $\mathbf{R}^n$. Vectors are *equivalent* (and we shall regard them as equal) if they have the same length and point in the same direction. We will use the same notation $(x_1, \ldots, x_n)$ to denote the point $(x_1, \ldots, x_n) \in \mathbf{R}^n$ and the vector that points from the origin to the point $(x_1, \ldots, x_n)$ in $\mathbf{R}^n$ (Fig. 1.6).

Vectors are added to and subtracted from each other in the usual way, by placing the vectors end to end (Fig. 1.4). Vectors can also be added to points. If P and Q are points in $\mathbf{E}^n$ and $\overrightarrow{v}$ is the vector that points from

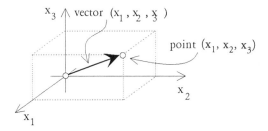

FIGURE 1.6. Vectors and points in $\mathbf{R}^3$.

P to Q then, as is shown in Figure 1.7,

$$Q = P + \overrightarrow{v}, P = Q - \overrightarrow{v}, \text{and}\, \overrightarrow{v} = \overrightarrow{PQ} = Q - P.$$

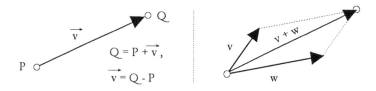

FIGURE 1.7. Adding vectors to points and vectors.

Orientation in the Plane

In the plane, reflections produce mirror images of objects, with the line in which one is reflecting serving as the mirror. Reflected writing has the same size and shape as the original, but it is still hard to read because its *orientation* has been changed. Here, "orientation" refers to a choice of direction of rotation – clockwise or counterclockwise in the plane. Reflections reverse orientation by changing clockwise rotations into counterclockwise rotations and vice-versa (Fig. 1.8).

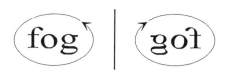

FIGURE 1.8. Reflections reverse orientation.

Orientation in Space

Orientation in $\mathbf{E}^3$ is determined by using the "right hand rule" (Fig. 1.9). Let $(\overrightarrow{A}, \overrightarrow{B}, \overrightarrow{C})$ be an ordered triple of nonzero vectors in $\mathbf{E}^3$, not all parallel to the same plane. $(\overrightarrow{A}, \overrightarrow{B}, \overrightarrow{C})$ is *positively oriented* if, when you point the thumb of your right hand in the $\overrightarrow{A}$ direction and your first finger in the $\overrightarrow{B}$ direction, then the rest of the fingers of your right hand curl toward the $\overrightarrow{C}$ direction. $(\overrightarrow{A}, \overrightarrow{B}, \overrightarrow{C})$ is *negatively oriented* if the rest of your fingers point in the $-\overrightarrow{C}$ direction.

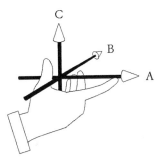

FIGURE 1.9. Right-hand rule.

Reflections in space reverse orientations by mapping positively oriented triples of vectors to negativly oriented triples and vice-versa.

1.5 Translations

Translations move objects along a straight line through space without rotating the objects or changing their orientation (Fig. 1.10). There is a one-to-one correspondence between vectors in $\mathbf{E}^n$ and translations: the translation T associated to a vector $\overrightarrow{v} \in \mathbf{E}^n$ acts by adding $\overrightarrow{v}$ to each point,

$$T(P) = P + \overrightarrow{v}.$$

$\overrightarrow{v}$ is the *displacement vector* for the translation T.

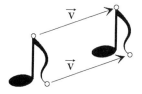

FIGURE 1.10. A translation.

In coordinates, if $\overrightarrow{v} = (v_1, \ldots, v_n)$ then

$$T(x_1, \ldots, x_n) = (x_1 + v_1, \ldots, x_n + v_n).$$

Exercise 1.5.1 Let S be the square whose vertices are $(0,0)$, $(1,0)$, $(1,1)$, and $(0,1)$, and let T be the translation associated to the vector $\overrightarrow{v} = (2,5)$. Find $T(S)$.

Composition of translations is equivalent to vector addition. If T_v is translation by $\overrightarrow{v}$ and T_w is translation by $\overrightarrow{w}$ then the composition $T_v \circ T_w$ is translation by the sum $\overrightarrow{v} + \overrightarrow{w}$:

$$T_v \circ T_w = T_{v+w}.$$

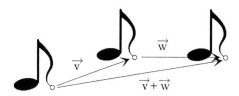

FIGURE 1.11. Composition of translations.

The inverse of a translation T_v also is a translation:

$$T_v^{-1} = T_{-v}. \tag{1.1}$$

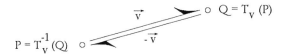

FIGURE 1.12. Inverse of a translation.

The next proposition says that every translation is a composition of reflections:

Proposition 1.5.1 *Let L_1 and L_2 be parallel lines in $\mathbf{E}^2$. Let $\overrightarrow{v}$ be the vector that points from L_1 to L_2 at right angles to the two lines, let f_1 be reflection in L_1, and let f_2 reflection in L_2. Then the composition $f_2 \circ f_1$ is translation by the vector $2\overrightarrow{v}$.*

Proof. (See Fig. 1.13). Let P be an arbitrary point. Let $\overrightarrow{A}$ and $\overrightarrow{B}$ be vectors perpendicular to the two lines, with $\overrightarrow{A}$ pointing from P to L_1 and $\overrightarrow{B}$ pointing from $f_1(P)$ to L_2. Then

$$P + \overrightarrow{A} \ \in \ L_1,$$

$$P + 2\overrightarrow{A} \;=\; f_1(P),$$
$$P + 2\overrightarrow{A} + \overrightarrow{B} \;\in\; L_2,$$
$$P + 2\overrightarrow{A} + 2\overrightarrow{B} \;=\; f_2(f_1(p)).$$

Because $\overrightarrow{A} + \overrightarrow{B}$ points from L_1 to L_2 and because all three vectors $\overrightarrow{v}$, $\overrightarrow{A}$ and $\overrightarrow{B}$ are perpendicular to L_1 and L_2, it follows that $\overrightarrow{v} = \overrightarrow{A} + \overrightarrow{B}$. Hence $f_2(f_1(P)) = P + 2\overrightarrow{v}$ for all P.

This completes the proof.

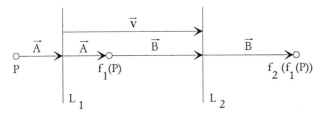

FIGURE 1.13.

A similar result with a similar proof holds for reflections in parallel planes in $\mathbf{E}^3$.

1.6   Rotations

Rotations in the plane

A rotation in the plane is performed by revolving the plane around a given point (the *center of the rotation*) through a given angle. Thus it takes two pieces of data to completely describe a rotation in $\mathbf{E}^2$, a point and an angle. Counterclockwise rotations sweep out positive angles; clockwise rotations sweep out negative angles.

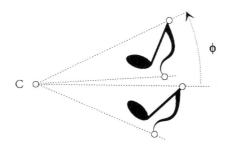

FIGURE 1.14. A rotation.

We will sometimes use the notation

$$R_{C,\phi} = (\text{the rotation with center } C \text{ and angle } \phi)$$

Rotation in the x, y plane through an angle ϕ about the origin O is given by the formula

$$R_{O,\phi}(x, y) = (x \cos \phi - y \sin \phi, x \sin \phi + y \cos \phi).$$

To see this, express (x, y) in polar coordinates (r, θ):

$$x = r \cos \theta, \quad y = r \sin \theta.$$

Then

$$
\begin{aligned}
R_{O,\phi}(x, y) &= (r \cos(\theta + \phi), r sin(\theta + \phi)) \\
&= (r(\cos \theta \cos \phi - \sin \theta \sin \phi), r(\cos \theta \sin \phi + \sin \theta \cos \phi)) \\
&= (x \cos \phi - y \sin \phi, x \sin \phi + y \cos \phi)
\end{aligned}
$$

by the addition formulas for trigonometric functions (Exercise 1.8.9).

The inverse of a rotation is a rotation with the same center and the opposite angle:

$$(R_{p,\phi})^{-1} = R_{p,-\phi}. \tag{1.2}$$

The next exercise shows that every rotation, just like every translation, is a composition of reflections.

Exercise 1.6.1 Let L_1, L_2 be two lines intersecting at a point P in $\mathbf{E}^2$. Let f_1 be reflection in L_1 and let f_2 be reflection in L_2. Show that the composition $f_2 \circ f_1$ is a rotation around P, and the angle of the rotation is equal to twice the angle formed by the two intersecting lines. (Hint: imitate the proof of Proposition 1.5.1).

Exercise 1.6.2 Find a formula for rotation through an angle ϕ around an arbitrary point $(a, b) \in \mathbf{R}^2$.

Rotations in Space
A rotation $R_{A,\phi}$ in three dimensional space has an angle ϕ and an axis A. A is a *directed line* – a line with a sense of direction. The direction of the axis determines the direction of positive rotation by using the "right hand rule": if the thumb of your right hand points in the direction of the axis then the fingers of your right hand curl towards positive rotations. See Fig. 1.9.

It is worth noticing that both translations and rotations preserve orientations. This follows from the fact that every translation and every rotation is the composition of two reflections (Proposition 1.5.1 and Exercise 1.6.1). Each of these reflections reverses orientation but the two orientation reversals cancel each other out, so in the end the orientation is the same as it was at the beginning.

1.7   Applications and Examples

Example 1.7.1 A boy intends to collect some water from a stream and then carry it home. He wants to find the shortest path that takes him to the stream and then to his house. What route should he take? (See Fig. 1.15.)

Obviously he should go straight to some point on the bank of the stream, fill his bucket, and then go straight home. To which point on the stream should he go?

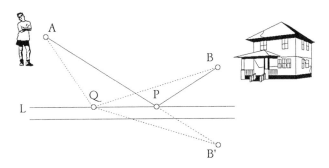

FIGURE 1.15. Shortest path?

Solution. Assume the boy is at A, his home is at B, and the bank of the stream forms a line L. Assume further that A and B are on the same side of L (since the solution is obvious if they are on opposite sides).

Claim 1.7.1 *Let B' be the reflection of B in L and let*

$$P = \overline{AB'} \cap L.$$

I claim that APB is the shortest path.[2]

Proof. It suffices to show that if $Q \neq P$ is any other point on L then $AP + PB < AQ + QB$.

Since reflections preserve distance, $QB = QB'$. Thus

$$AQ + QB = AQ + QB'.$$

Likewise $PB = PB'$, so

$$AP + PB = AP + PB'.$$

Clearly

$$AP + PB' < AQ + QB',$$

---

[2]Notation: If $A_1, A_2, \ldots, A_n$ are a sequence of points then $A_1 A_2 \ldots A_n$ is a union of line segments $A_1 A_2 \ldots A_n = \overline{A_1 A_2} \cup \overline{A_2 A_3} \cup \ldots \cup \overline{A_{n-1} A_n}$.

because A, P, and B' lie in a straight line while A, Q, and B' do not. Hence

$$AP + PB < AQ + QB$$

which proves the claim.

Exercise 1.7.1 Let L be a line in the plane, B a point not on L, B' the reflection of B in L, and $Q = \overline{BB'} \cap L$. Prove that Q is the closest point to B on L.

Exercise 1.7.2 Suppose that in Example 1.7.1 $A = (0, 2)$, $B = (6, 1)$, and L is the x axis. Where is P?

When a light ray bounces off a flat mirror the *angle of incidence,* the angle between the incoming ray and a line perpendicular to the mirror, is equal to the *angle of reflection,* the angle between the outgoing ray and a line perpendicular to the mirror (Fig. 1.16).

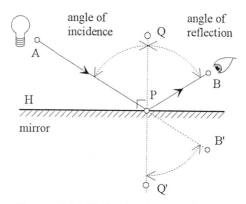

FIGURE 1.16. Reflection from a mirror.

To bounce a light beam from a point A to a point B on the same side of a planar mirror one should aim the light toward the reflection B' of B in the plane of the mirror. To see why, let H be the plane of the mirror, B' the reflection of B in H, and $P = \overline{AB'} \cap H$. Let Q be a point on the same side of H as A, with $\overleftrightarrow{QP}$ perpendicular to H, and let Q' be the reflection of Q in H.

$$\angle BPQ = \angle B'PQ'$$

because reflections preserve angles, while

$$\angle B'PQ' = \angle APQ$$

because $\angle B'PQ'$ and $\angle APQ$ are "vertical angles" (a $180°$ rotation about P in the plane $\overline{APB}$ takes $\angle B'PQ'$ to $\angle APQ$). Hence $\angle APQ = \angle BPQ$, so light travels along the path APB.

Remark. From our discussion of Example 1.7.1 it follows that APB is the shortest path from A to the mirror and then to B. This is an example of Hamilton's "Principle of Least Action," an important principle in physics that says that all types of waves, including strange things like "matter waves", tend to follow paths that minimize some physical quantity ("action"). The interested reader should consult Richard Feynman's famous little book QED [8] for a simple and intuitive discussion of the role of Hamilton's principle in quantum mechanics.

Example 1.7.2 Two mirrors meet in a $70°$ angle (Fig. 1.17). Someone stands between the mirrors with her eye at point B searching for images of an object that is located at point A. How many images of the object will she see? In what direction and how far away will they appear to be?

For simplicity's sake assume that A and B lie in a plane H that is perpendicular to the line where the two mirrors intersect. This enables us to treat the problem as a problem in plane geometry: since H is perpendicular to both mirrors a light ray reflecting from A to B will remain in H for its entire trip.

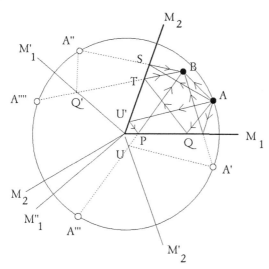

FIGURE 1.17. Multiple reflections.

Fig. 1.17 shows five images of A and their apparent locations as they appear from an observer at B. A itself is one of the "images" since the observer could look directly at A. Another image is formed by light that goes from A to the first mirror, M_1, reflects, and then goes to B. This image appears to be located at the reflection A' of A in M_1, because the reflected light comes from the same direction as A' and it travels a distance equal to $A'B$. Similarly light traveling along the path ASB forms an image that appears to be located at A'', the reflection of A in M_2.

The image A'''' is formed as follows. M_1' is the reflection of the mirror M_1 in M_2. A'''' is the reflection of A'' in M_1'. $\overline{A''''B}$ intersects M_1' at Q' and M_2 at T. Reflecting $\overline{A''''Q'}$ in M_1' one gets $\overline{A''Q'}$, then reflecting the path $A''Q'T$ in M_2 one obtains the path AQT. Thus light reflects from A to B along the path $AQTB$, and the image appears to be located at A''''.

Similarly, light reflects along the path $AU'PB$ producing an image at A'''. A''' is the reflection of A' in M_2', while M_2' is the reflection M_2 in M_1. $U = \overline{A'''B} \cap M_2'$, $P = \overline{A'''B} \cap M_1$, $\overline{A'''U}$ reflects to $\overline{A'U}$ in M_2', and $A'UP$ reflects to $AU'P$ in M_1.

One can construct all of these reflections by folding the plane over (see page 4). For instance the path $AU'PB$ results from folding $\overline{A'''B}$ once over M_2' and then over M_1. Or one could think of $\overline{A'''B}$ as the result of unfolding $AU'PB$, once over M_1 and then over M_2'.

This folding construction makes it easy to see why no reflections are possible in this example except the five that were constructed above. When one unfolds a path, the points where the light reflects (e.g. U' and P in $AU'PB$) become points where the unfolded line crosses a mirror (e.g. U and P in $\overline{A'''B}$). As the unfolding progresses the successive images of A (A',A''' in the unfolding of $AU'PB$) all lie on the same side of the line that connects B to the point where the two mirrors intersect. No more images are possible because it is impossible to produce any more images of A without passing from one side of this line to the other.

The number of reflections, and hence the number of images seen by the observer, varies in different examples. It depends mainly on the angle between the two mirrors, and also is affected somewhat by the position of the viewer (B) and the position of the object (A) relative to the mirrors.

Exercise 1.7.3 Find all possible paths that light could take in Example 1.7.2 if the angle between the mirrors were

a) $90°$.

b) $60°$.

c) $50°$.

d) Can you find a formula for the number of paths from A to B as a function of the angle between the mirrors and the positions of the points A and B?

Exercise 1.7.4 What modifications should be made in the discussion of Example 1.7.2 if A and B do not lie in a plane that is perpendicular to the intersection of the two mirrors?

Exercise 1.7.5 Balls bounce the same way as light if you neglect the effects of friction. Suppose a frictionless ball is placed at the center of the pool table that is twice as long as it is wide (Fig. 1.18).

a) Show how to make a shot that will send the ball into the corner pocket C, bouncing it off each side of the table at least once along the way (minimum of four bounces). Neglect friction. (Hint: reflect the pool table repeatedly over each of its sides to generate a grid pattern like the one in Fig. 1.18, then construct your path by folding up the appropriate straight line. Fig. 1.18 shows a shot that lands the ball in the pocket C after three bounces.)

b) Prove that the shot called for in part a) can be done with five bounces, but it cannot be done with only four.

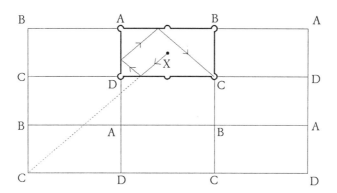

FIGURE 1.18. A pool shot.

Example. (Adapted from [22, page 10 problem 2]). A road is proposed that will connect two towns A and B on opposite sides of a river. The road will cross the river in a bridge that is perpendicular to the riverbanks. Where should the bridge be placed so as to minimize the total length of the road? (See Fig. 1.19).

Solution. Let $\vec{v}$ be the vector that points from the riverbank on the A side of river to the riverbank on the B side, perpendicular to the river. Set $A' = A + \vec{v}$. The bridge should be built at the point P' where $\overline{A'B}$ crosses the riverbank on the B side of the river.

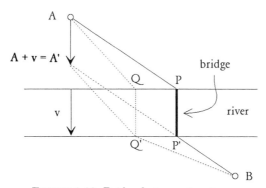

FIGURE 1.19. Bridge between two towns.

Proof. We must show that if one puts the bridge anywhere else he will get a longer road. A bridge at P' extends from P' to the point $P = P' - \vec{v}$

on the opposite riverbank. With the bridge at P', the length of the road is

$$AP + |\overrightarrow{v}| + P'B$$

(Notation: $|\overrightarrow{v}|$ stands for the length of the vector $\overrightarrow{v}$).

Let $Q' \neq P'$ be any other point on the B side of the river and let $Q = Q' - \overrightarrow{v}$ be the corresponding point on the opposite riverbank. If the bridge were built at Q' then the road's length would be

$$AQ + |\overrightarrow{v}| + Q'B.$$

Translation by $\overrightarrow{v}$ maps $\overline{AP}$ to $\overline{A'P'}$ and $\overline{AQ}$ to $\overline{A'Q'}$. Thus

$$AP + |\overrightarrow{v}| + P'B = A'P' + |\overrightarrow{v}| + P'B$$

and

$$AQ + |\overrightarrow{v}| + Q'P = A'Q' + |\overrightarrow{v}| + Q'P.$$

Clearly

$$A'P' + P'B < A'Q' + Q'B$$

since the points A', P', and B lie on a line. Hence $A'P' + |\overrightarrow{v}| + P'B < A'Q' + |\overrightarrow{v}| + Q'P$, so $APP'B$ is the shortest road.

Exercise 1.7.6 Show how to find the shortest road between two towns separated by an arbitrary number of rivers that must be bridged (Fig. 1.20). Prove that your road really is the shortest possible.

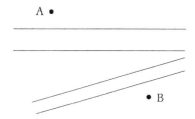

FIGURE 1.20. Two rivers between two towns.

Example 1.7.3 Let $\triangle ABC$ be an arbitrary triangle in the plane, and let $\triangle A'BC$, $\triangle AB'C$, $\triangle ABC'$ be equilateral triangles attached to the outside of $\triangle ABC$. Then

$$AA' = BB' = CC'.$$

(See Fig. 1.21).

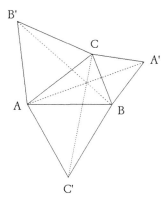

FIGURE 1.21. Equilateral triangles on a triangle.

Proof. A $60°$ rotation about the point C carries B to A' and B' to A simultaneously. Since rotations preserve distances, it follows that $BB' = A'A$. A similar argument (rotate by $60°$ around B) shows that $CC' = A'A$.

It turns out (see Exercise 1.9.2) that the three lines $\overleftrightarrow{AA'}$, $\overleftrightarrow{BB'}$, $\overleftrightarrow{CC'}$ in Example 1.7.3 all intersect at the same point, in $60°$ angles. The point D where they intersect is the solution to another minimization problem: if none of the angles of $\triangle ABC$ is larger than $120°$ then the sum of the distances $AP + BP + CP$ is smallest when $P = D$. For an elementary proof see [4, Chap. VII §5].

1.8   Some Key Results of High School Geometry: The Parallel Postulate, Angles of a Triangle, Similar Triangles, and the Pythagorean Theorem

Two lines in $\mathbf{E}^n$ are *parallel* if and only if they are equal or they lie in a common plane and do not intersect.

The Parallel Postulate: Given a line L and a point P in $\mathbf{E}^n$ there is exactly one line through P that is parallel to L.

In fact if T is a translation that takes some point on L to P then $T(L)$ is parallel to L (Fig. 1.22).

The parallel postulate was the most controversial of Euclid's postulates for geometry. Many scholars felt that it should be possible to deduce the parallel postulate from Euclid's other postulates. Although it was later

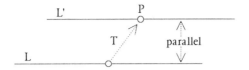

FIGURE 1.22. Parallel postulate.

proved to be impossible to deduce the parallel postulate from the other postulates, efforts to do so led to the invention of various "non-Euclidean geometries" in which the parallel postulate is violated. Besides being interesting to mathematicians, some non-Euclidean geometries have found practical application, the most famous being in Einstein's General Theory of Relativity.

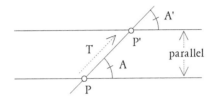

FIGURE 1.23. Alternate Interior-Exterior Angles.

Recall that two figures A and B are congruent (in symbols $A \cong B$) if there is an isometry T such that $T(A) = B$.

Alternate interior-exterior angles $\angle A$ and $\angle A'$ are formed when parallel lines intersect a transversal (Fig. 1.23). Alternate interior-exterior angles are congruent since a translation maps one of them to the other.

Conversely, if $\angle A \cong \angle A'$ then L and L' are parallel. To see this let L'' be the line through P' that is parallel to L. $\angle A$ and one of the angles $\angle A''$ that are formed by L'' and the transversal are alternate interior-exterior angles. Hence $\angle A'' \cong \angle A'$. Since $\angle A''$ and $\angle A'$ have the same vertex it follows that they are the same angle, so $L'' = L'$. Hence L' is parallel to L.

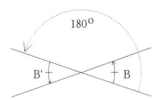

FIGURE 1.24. Vertical Angles.

Two pairs of *vertical angles* are formed when two lines intersect (Fig. 1.24). Vertical angles are congruent since a 180° rotation about the point of intersection maps one angle to the other.

Theorem 1.8.1 *The sum of the angles of a Euclidean triangle is* $180°$.

Proof. Given $\triangle ABC$, draw a line L through A parallel to $\overline{BC}$. Label angles as in Fig. 1.25. $\angle A$ and $\angle A'$ are vertical angles so $\angle A \cong \angle A'$, while $\angle B \cong \angle B'$ and $\angle C \cong \angle C'$ because they are alternate interior-exterior angles. Clearly $\angle A' + \angle B' + \angle C' = 180°$.

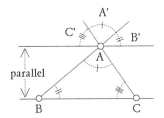

FIGURE 1.25. Sum of angles $= 180°$.

Theorem 1.8.1 depends strongly on the parallel postulate–it is *not* true in non-Euclidean geometries where "space" is "curved". The difference between the sum of the angles of a triangle and $180°$ is a measure of the curvature of space; if the difference is not zero the space is curved rather than flat. In this sense our own physical universe is a curved space: if you build a very large triangle by joining three vertices together with curves of minimal length then the angles of the triangle generally will not add up to $180°$. The General Theory of Relativity explains that this bending of space is a manifestation of gravitation.

Many key results from high school geometry follow from elementary facts about area. Often the area inside a figure can be computed by cutting the figure up and rearranging the pieces until one obtains a figure of known area. For instance, starting with the fact that the area of a rectangle is the length of its base times its height, one finds that:

$$(\text{area of a parallelogram}) = (\text{base}) \times (\text{height}).$$

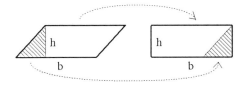

FIGURE 1.26. Area of a parallelogram.

$$\text{(area of a triangle)} = \frac{1}{2}\text{(base)} \times \text{(height)}.$$

FIGURE 1.27. Area of a triangle.

$$\text{(area of a circle)} = \frac{1}{2}\text{(circumference)} \times \text{(radius)}.$$

(If r is its radius then the area of the circle is πr^2 since its circumference is $2\pi r$).

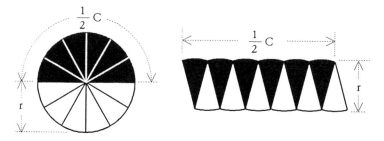

FIGURE 1.28. Area of a circle.

To obtain the last formula cut the circle into an arbitrarily large number of infinitesimally thin sectors and rearrange them into an approximate parallelogram as shown in Fig. 1.28. Then take the limit as the number of sectors approaches infinity.

Proposition 1.8.1 *If L is parallel to $\overline{AB}$ and $C, C' \in L$, then $\triangle ABC$ and $\triangle A'B'C'$ have the same area (Fig. 1.29).*

Proof. Both triangles have the same base and height.

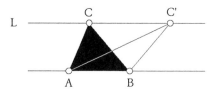

FIGURE 1.29. The triangles have equal area.

A similar result holds for parallelograms (Fig. 1.30).

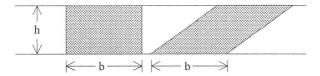

FIGURE 1.30. The parallelograms have equal area.

Definition 1.8.1 Similar Triangles. Triangles $\triangle ABC$ and $\triangle A'B'C'$ are *similar* if $\angle A \cong \angle A'$, $\angle B \cong \angle B'$, and $\angle C \cong \angle C'$.

The next result is often stated as an axiom in high school texts.

Theorem 1.8.2 Similar Triangles are Proportional.
If $\triangle ABC$ and $\triangle A'B'C'$ are similar then

$$\frac{A'B'}{AB} = \frac{A'C'}{AC} = \frac{B'C'}{BC}.$$

Proof. (Adapted from Euclid [6, Book VI, Proposition 2]). Without loss of generality, we may assume that $AB \geq A'B'$. Since $\angle A \cong \angle A'$ there exists an isometry T such that $T(\overrightarrow{A'B'}) = \overrightarrow{AB}$ and $T(\overrightarrow{A'C'}) = \overrightarrow{AC}$. By applying T we may arrange the triangles as in Fig. 1.31, where $A = A'$, $B' \in \overline{AB}$, and $C' \in \overline{AC}$.

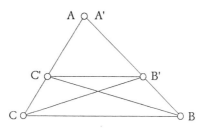

FIGURE 1.31.

Regard $\triangle ABC'$ and $\triangle A'B'C'$ as triangles with their bases on $\overleftrightarrow{AB}$ and their heights equal to the distance from $\overleftrightarrow{AB}$ to C'. The formula for the area of a triangle says that

$$\frac{A'B'}{AB} = \frac{\text{area}(\triangle A'B'C')}{\text{area}(\triangle ABC')}. \tag{1.3}$$

Similarly,

$$\frac{A'C'}{AC} = \frac{\text{area}(\triangle A'C'B')}{\text{area}(\triangle ACB')}. \tag{1.4}$$

Now $\overleftrightarrow{B'C'}$ is parallel to $\overleftrightarrow{BC}$ since $\angle C \cong \angle C'$, so the distance from B to the line $\overleftrightarrow{B'C'}$ equals the distance from C to $\overleftrightarrow{B'C'}$. Hence

$$\text{area}(\triangle B'C'C) = \text{area}(\triangle B'C'B). \tag{1.5}$$

But
$$\text{area}(\triangle ABC') = \text{area}(\triangle A'B'C') + \text{area}(\triangle B'C'B)$$
and
$$\text{area}(\triangle ACB') = \text{area}(\triangle A'B'C') + \text{area}(\triangle B'C'C).$$

So by Equation 1.5,
$$\text{area}(\triangle ABC') = \text{area}(\triangle ACB').$$

Plug this result into Equations 1.3 and 1.4 and get
$$\frac{A'B'}{AB} = \frac{A'C'}{AC}.$$

An analogous argument shows that $A'B'/AB = B'C'/BC$.
This completes the proof.

Remark. From the law of cosines (Exercise 1.8.6) it follows that the converse to Theorem 1.8.2 is also true: if $A'B'/AB = A'C'/AC = B'C'/BC$ then $\triangle A'B'C'$ is similar to $\triangle ABC$ (Exercise 1.8.7).

The Pythagorean Theorem

The following generalization of the Pythagorean Theorem was proved by Pappus of Alexandria in the fourth century, A.D. (See [7, Lecture 4] for further discussion and generalizations.)

Theorem 1.8.3 *Let $\triangle ABC$ be an arbitrary triangle in $\mathbf{E}^2$, not necessarily a right triangle. Erect parallelograms $ABDE$ and $ACFG$ on the outside of $\triangle ABC$ so that $ABDE$ meets $\triangle ABC$ along the edge $\overline{AB}$ and $ACFG$ meets $\triangle ABC$ along the edge $\overline{AC}$. Let P be the point where the lines $\overleftrightarrow{DE}$ and $\overleftrightarrow{FG}$ meet. Erect a third parallelogram $BCHI$ on the outside of $\triangle ABC$ so that the vectors*
$$\overrightarrow{PA} = \overrightarrow{BI}$$
are equal. Then
$$area(ABDE) + area(ACFG) = area(BCHI).$$

(See Fig. 1.32).

Proof. Let $Q = \overleftrightarrow{DE} \cap \overleftrightarrow{BI}$, $R = \overleftrightarrow{FG} \cap \overleftrightarrow{CH}$, $J = \overleftrightarrow{PA} \cap \overleftrightarrow{BC}$, and $K = \overleftrightarrow{PA} \cap \overleftrightarrow{IH}$. $\overline{QB}$, $\overline{PA}$, $\overline{RC}$, and $\overline{JK}$ are parallel. Two parallelograms with the same base and height have the same area, so
$$area(ABDE) = area(ABQP) = area(JKIB),$$
and
$$area(ACFG) = area(ACRP) = area(JKHC).$$

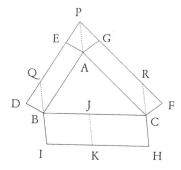

FIGURE 1.32. The Generalized Pythagorean Theorem.

Clearly,

$$\text{area}(BCHI) = \text{area}(JKIB) + \text{area}(JKHC),$$

so the proof is complete.

Exercise 1.8.1 The Pythagorean theorem. Deduce the Pythagorean theorem from Theorem 1.8.3: if $\angle A = 90^\circ$ then $(AB)^2 + (AC)^2 = (BC)^2$.

The theorem on similar triangles (Theorem 1.8.2) and the Pythagorean theorem form the basic link between geometry and algebra. The "point-slope" equation of a line is a consequence of the theorem on similar triangles: if (x_1, y_1) and (x_2, y_2) are two points on the line and $m = (y_2 - y_1)/(x_2 - x_1)$ then, given any other point (x, y) on the line, the ratios $(y - y_1)/(x - x_1)$ and $(y_2 - y_1)/(x_2 - x_1)$ are equal (by similar triangles). Hence $(y - y_1) = m(x - x_1)$. It follows from this that equations of the form $Ax + By + C = 0$ represent lines. Equations of the form $Ax^2 + Bxy + Cy^2 + Dx + Ey + F = 0$ represent circles, ellipses, hyperbolas, and parabolas because of the Pythagorean theorem, which says that the square of the distance between (x_1, y_1) and (x_2, y_2) is $d^2 = (x_2 - x_1)^2 + (y_2 - y_1)^2$.

This marriage of geometry and algebra has been exceptionally fruitful. It enables us to apply the intuitions of geometry to algebraic problems and the precision of algebra to geometric problems, producing an enormous increase in the depth and scope of both fields.

Similar triangles and the Pythagorean theorem also provide the foundation of trigonometry and much of the science of measurement. Using similar triangles one can recover the dimensions of an object by measuring a scale model of the object. Each angle on the object is the same as a corresponding angle on the model, so each length on the object differs from the corresponding length on the model by the same scale factor. Therefore to obtain a distance on the original object one need only multiply the corresponding length on the model by this scale factor.

Example. How high is your science building? From two points A and B on the ground (see Fig. 1.33) I could see point C at the top of our science building. Using a protractor and a tape measure, I found that the angle of elevation from point A to point C is $\angle A = 19°$, the angle of elevation from B to C is $\angle B = 43°$ and the distance from A to B is $AB = 39$ft. The measurements were taken at my eye level, about 5.5 ft. above the ground.

To find the height of the building I made a scale drawing of $\triangle ABC$. I drew a line segment $\overline{A'B'}$ of some convenient length on a piece of paper (I used $A'B' = 19.5$ cm. but any other length would do just as well). Then I drew a line through A' meeting $\overline{A'B'}$ in a $19°$ angle and another line through B' meeting $\overline{A'B'}$ in a $43°$ angle. Let C' be the point where the two lines meet. Drop a perpendicular $\overline{C'D'}$ from C' to the line $\overleftrightarrow{A'B'}$. $\overline{C'D'}$ represents the side $\overline{CD}$ of the science building (Fig. 1.33). Measuring my drawing with a ruler I found that

$$C'D' \approx 10.9 \text{ cm.}$$

The scale factor in my drawing is the ratio $r = AB/A'B'$ of the true length AB to the distance $A'B'$ between the corresponding points in the drawing. In my case $AB = 39$ ft. and $A'B' = 19.5$ cm. so the scale factor is $r = 39$ ft./19.5 cm.. Hence the true distance CD is

$$\begin{aligned} CD \;&=\; C'D' \times \frac{39 \text{ ft.}}{19.5 \text{ cm.}} \\ &\approx\; 10.9 \text{ cm.} \times \frac{39 \text{ ft.}}{19.5 \text{ cm.}} \\ &\approx\; 21.8 \text{ ft.} \end{aligned}$$

My eye level is about 5.5 ft. above the ground, so it follows that our science building is approximately 21.8 ft. + 5.5 ft. $\approx$ 27.3 ft. high.

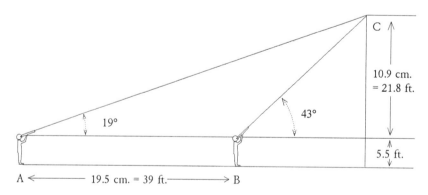

FIGURE 1.33. Measurement of a remote object.

Figure 1.34 shows the simple device that I used to measure the angles of elevation. To make it you need a straight stick, a protractor, a string, some washers for a weight, and some tape, and two people: one to sight the object and one to read the scale.

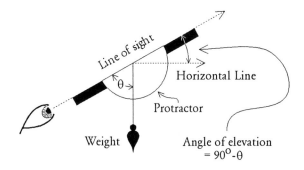

FIGURE 1.34. A device for measuring angle of elevation.

Exercise 1.8.2 Make a scale drawing and use it to find the height of the library building on your campus.

Exercise 1.8.3 From two points A and B on their side of a river, a group of sightseers can see a dock C on the opposite bank. Using a protractor and a tape measure, they find that $\angle CAB = 75^\circ$, $\angle ABC = 80^\circ$, and $AB = 100$ft. Make a scale drawing and use it to find the width of the river.

Long ago people realized that for solving this type of problem it would be useful to have a table of triangles and their dimensions. Instead of having to draw and measure scale models of triangles one could simply look up their measurements in the table. In fact it is enough simply to have a table of right triangles, because every triangle can be broken down into two right triangles (Fig. 1.35).

FIGURE 1.35. Every triangle is made from two right triangles.

A table of trigonometric functions *is* such a table; sines and cosines are the lengths of the legs of a right triangle if the length of its hypotenuse is one. In practice one usually can achieve better results using a table of trigonometric functions than one can by drawing and measuring because it is difficult to draw and measure really accurately, whereas trigonometric functions can be computed to any desired precision.[3] Nevertheless, trigono-metric tables are in essence nothing but tables of similar triangles, a sub-stitute for scale drawings.

Exercise 1.8.4 Use a table of trigonometric functions (or a calculator) instead of a scale drawing to work Exercise 1.8.3. Compare the results.

Exercise 1.8.5 The Law of Sines
Given a triangle with a side of length a opposite vertex A, side b opposite vertex B, and side c opposite vertex C, prove that

$$\frac{a}{\sin \angle A} = \frac{b}{\sin \angle B} = \frac{c}{\sin \angle C}.$$

(Hint: Compute the area of the triangle three times using sides a, b, and c, respectively, as the base. For a sharper version of the Law of Sines see Exercise 1.9.11.)

Exercise 1.8.6 The Law of Cosines.
Given a triangle with sides a, b, and c, and vertex C opposite side c, prove that

$$c^2 = a^2 + b^2 - 2ab \cos(\angle C).$$

(Hint: break the triangle up into two right triangles as in Fig. 1.35 and apply the Pythagorean theorem).

Exercise 1.8.7 Use the law of cosines (Exercise 1.8.6) to prove the con-verse to Theorem 1.8.2: if $A'B'/AB = A'C'/AC = B'C'/BC$ then $\triangle A'B'C'$ is similar to $\triangle ABC$.

Exercise 1.8.8 The Dot Product. The *dot product* of two vectors $\overrightarrow{v} = (x_1, y_1)$, and $\overrightarrow{w} = (x_2, y_2)$ in $\mathbf{R}^2$ is

$$(x_1, y_1) \cdot (x_2, y_2) = x_1 x_2 + y_1 y_2.$$

Use the law of cosines (Exercise 1.8.6) and the Pythagorean theorem to prove that
   a) $|\overrightarrow{v}|^2 = \overrightarrow{v} \cdot \overrightarrow{v}$, where $|\overrightarrow{v}|$ is the length of $\overrightarrow{v}$,
   b) $\overrightarrow{v} \cdot \overrightarrow{w} = |\overrightarrow{v}||\overrightarrow{w}| \cos \theta$, where θ is the angle between $\overrightarrow{v}$ and $\overrightarrow{w}$.
   (Hint: Apply the law of cosines to the triangle whose sides are $\overrightarrow{v}$, $\overrightarrow{w}$, and $\overrightarrow{v} - \overrightarrow{w}$.)

---

[3]For instance, using the Taylor series

$$\cos x = 1 - \frac{1}{2}x^2 + \frac{1}{4!}x^4 - \frac{1}{6!}x^6 + \cdots$$

$$\sin x = x - \frac{1}{3!}x^3 + \frac{1}{5!}x^5 - \frac{1}{7!}x^7 + \cdots.$$

More efficient methods for machine computation are discussed in [19].

Exercise 1.8.9 Addition Formulas.

a) By applying the law of cosines to the triangle in Fig. 1.36, deduce that

$$\cos(\phi + \theta) = \cos(\phi)\cos(\theta) - \sin(\phi)\sin(\theta)$$

and

$$\sin(\phi + \theta) = \sin(\phi)\cos(\theta) + \cos(\phi)\sin(\theta)$$

whenever $0 \le \phi, \theta < 90°$.

b) Using the identities $\cos(\theta) = \cos(-\theta) = -\cos(\theta + 180°)$ and $\sin(\theta) = -\sin(-\theta) = -\sin(\theta + 180°)$, generalize the results in part a) to all angles $-\infty < \phi, \theta < \infty$.

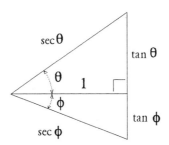

FIGURE 1.36. Addition formulas.

1.9 SSS, ASA, and SAS

Much of school geometry is taken up with the study of elementary relations ("angle-side-angle", "side-side-side", etc.) between congruent triangles. These relations, as well as all the rest of Euclidean geometry, follow from properties of isometries.

Proposition 1.9.1 Side-Angle-Side (SAS).

If $AB = A'B'$, $BC = B'C'$, and $\angle B \cong \angle B'$ then $\triangle ABC \cong \triangle A'B'C'$ see (Fig. 1.37.

FIGURE 1.37. Side-angle-side.

Proof. (Adapted from Euclid [6, Book I, Proposition 4]). Since $\angle B$ is congruent to $\angle B'$ there is an isometry T such that $T(\angle B) = \angle B'$. By composing it with a reflection if necessary we may assume that T takes $\overrightarrow{BA}$ to $\overrightarrow{B'A'}$ and $\overrightarrow{BC}$ to $\overrightarrow{B'C'}$. Because $AB = A'B'$ and $BC = B'C'$ it follows that $T(A) = A'$ and $T(C) = C'$, since isometries preserve distances. Therefore T maps $\triangle ABC$ to $\triangle A'B'C'$.

Proposition 1.9.2 Angle-Side-Angle (ASA).
If $\angle A \cong \angle A'$, $\angle B \cong \angle B'$, and $AB = A'B'$ then $\triangle ABC \cong \triangle A'B'C'$.

Proof. $\angle C \cong \angle C' \cong 180° - \angle A - \angle B$. By the law of sines (Exercise 1.8.5),
$$BC = \frac{(AB)(\sin A)}{\sin C} = \frac{(A'B')(\sin A')}{\sin C'} = B'C'.$$
Thus by SAS (Prop. 1.9.1) the triangles are congruent.

Proposition 1.9.3 Side-Side-Side (SSS).
If $AB = A'B'$, $AC = A'C'$, and $BC = B'C'$ then $\triangle ABC \cong \triangle A'B'C'$.

Proof. By the law of cosines (Exercise 1.8.6)
$$\begin{aligned}
\cos C &= \frac{(AC)^2 + (BC)^2 - (AB)^2}{2(AC)(BC)} \\
&= \frac{(A'C')^2 + (B'C')^2 - (A'B')^2}{2(A'C')(B'C')} = \cos C'.
\end{aligned}$$

Hence $\angle C \cong \angle C'$, so by SAS the triangles are congruent.

Isosceles Triangles and Arcs on a Circle

Corollary 1.9.1 Isosceles triangles.
$\angle A \cong \angle B$ in $\triangle ABC$ if and only if $AC = BC$.

Proof. If $\angle A \cong \angle B$ then by ASA there is an isometry T such that $T(A) = B$, $T(B) = A$, and $T(C) = C$. Thus $T(\overline{AC}) = \overline{BC}$, so $AC = BC$.
Conversely, if $AC = BC$ then by SSS there is an isometry T such that $T(A) = B$, $T(B) = A$, and $T(C) = C$, so $\angle A \cong \angle B$.

If $AC = BC$ and M is the midpoint of AB, then $\overleftrightarrow{CM}$ is the perpendicular bisector of $\overline{AB}$. For $\triangle CMA$ is congruent to $\triangle CMB$ by SSS, and $\angle CMA + \angle CMB = 180°$. This principle is used in an old-fashioned design for a carpenter's level (Fig. 1.38). Build a triangle $\triangle ABC$ with $AC = BC$. Mark the midpoint M of $\overline{AB}$ and hang a weight on a string attached at C.
To use the level rotate the triangle until the string crosses $\overline{AB}$ at M. Then $\overleftrightarrow{AB}$ will be a horizontal line.
According to Thompson, [21, page 100] this type of level was used in the eighteenth century for surveying the Mason-Dixon Line.
Another application of isosceles triangles is the following useful result from elementary geometry.

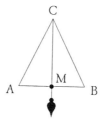

FIGURE 1.38. A Carpenter's Level.

Proposition 1.9.4 *Let C be a circle with center at P, A an arc of C, and $Q \in C$ a point not on A. Let R and S be the endpoints of A. Then*

$$\angle SPR = 2\angle SQR$$

(see Fig. 1.39).

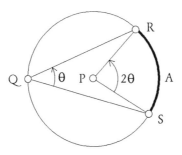

FIGURE 1.39.

Proof. (See Fig. 1.40). Let Q' be the opposite end of the diameter through Q. $PQ = PR$, so $\triangle QPR$ is isosceles. Hence $\angle QRP = \angle PQR$. Since the sum of the angles of $\triangle QRP$ is $180°$ it follows that

$$\angle RPQ = 180° - 2\angle PQR,$$

and since $\angle Q'PR = 180° - \angle RPQ$ we have

$$\angle Q'PR = 2\angle PQR.$$

Similarly,

$$\angle SPQ' = 2\angle SQP.$$

Since

$$\angle SPR = \angle SPQ' \pm \angle Q'PR$$

and

$$\angle SQR = \angle SQP \pm \angle PQR$$

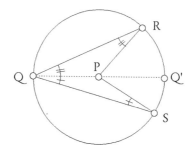

FIGURE 1.40.

with the same plus-or-minus sign in each case, it follows that $\angle SPR = 2\angle SQR$.

Corollary 1.9.2 *If R and S are opposite ends of a diameter of a circle and Q is any other point on the circle then $\angle SQR = 90°$.*

Exercise 1.9.1 (Adapted from [16, page 6]). The best seat in a certain theater is the seat marked "A". Find all other points from which the stage subtends the same angle as it does for the viewer in seat A (Fig. 1.41).

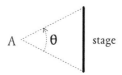

FIGURE 1.41.

Exercise 1.9.2 Show that the lines $\overleftrightarrow{AA'}$, $\overleftrightarrow{BB'}$, $\overleftrightarrow{CC'}$ in Example 1.7.3 all intersect in 60° angles at a single point, D. (Hint: let $D = \overleftrightarrow{AA'} \cap \overleftrightarrow{BB'}$. Show that $\angle C'DB = \angle BDA' = \angle A'DC = 60°$, then deduce that C', D, and C must be collinear).

Exercise 1.9.3 The navigator of a ship S saw landmarks in the distance at three points A, B, and C. Taking sightings from the deck of the ship she found that $\angle ASB = 100°$, $\angle BSC = 125°$, and $\angle CSA = 135°$. Then she located the points A, B, and C on a map and used Proposition 1.9.4 to find the exact position of her ship. How did she do it? (Fig. 1.42).

Exercise 1.9.4 Tangents to circles and spheres.
   a) A line is tangent to a circle if and only if it intersects the circle in exactly one point. Let C be a circle with center Z, $A \in C$, and L a line

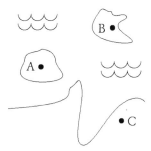

FIGURE 1.42. Sightings from a ship.

through A. Prove that L is tangent to C if and only if L is perpendicular to $\overline{ZA}$. (Hint: use the Pythagorean theorem).

b) State and prove a corresponding result for planes that are tangent to spheres.

Exercise 1.9.5 Tangents to circles (continued).

a) In some applications (e.g. drafting) it is useful to be able to construct a line through a given point P, tangent to a given circle C. One may try to do this by eye but it is hard to get it just right.

Here is a better way. Let Z be the center of C and M the midpoint of $\overline{ZP}$. Draw the circle with radius ZM and center M, and let A and B be the points where it intersects C.

Prove that $\overleftrightarrow{AP}$ and $\overleftrightarrow{BP}$ are tangent to C (Fig. 1.43).

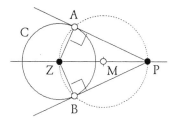

FIGURE 1.43. Tangents to a Circle.

b) Here is a way to draw lines tangent to two given circles. (This technique is used, for example, in drawing pictures of belts passing over pulleys). Start with two circles C,C', with radii r,r' and centers Z,Z', respectively.

Assume $r > r'$. To draw an "external" tangent (Fig. 1.44), first draw an auxiliary circle K with center Z and radius $r - r'$. With the technique from part a), locate a point $A \in K$ such that $\overleftrightarrow{AZ'}$ is tangent to K. Let B be the point where $\overrightarrow{ZA}$ intersects C. Starting at Z', draw a ray parallel to $\overrightarrow{ZA}$ and let B' be the point where it intersects C'. Prove that BB' is tangent to both C and C'.

To draw an "internal" tangent (Fig. 1.45) follow the same construction, except let $r+r'$ be the radius of K and make $\overrightarrow{ZA}$ and $\overrightarrow{Z'B'}$ point in opposite directions.

c) A belt wraps around two pulleys, which are mounted with their centers 4 feet apart. If the radius of one pulley is 1 ft. and the radius of the other is 2 ft., how long is the belt? (Assume the belt has zero thickness and does not cross itself).

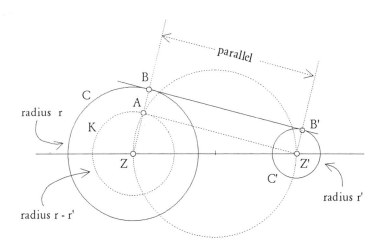

FIGURE 1.44. "External" Tangent to Two Circles.

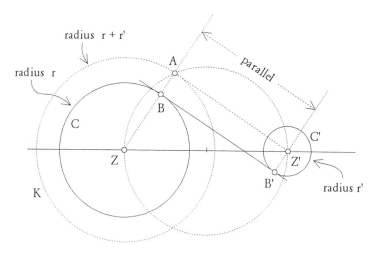

FIGURE 1.45. "Internal" Tangent.

Exercise 1.9.6 a) Prove: If $\overleftrightarrow{AP}$ and $\overleftrightarrow{BP}$ are tangent to a circle at distinct points A and B, respectively, then $AP = BP$ and the bisector of $\angle APB$ passes through the center of the circle.

b) State and prove a corresponding result for lines tangent to spheres.

Exercise 1.9.7 The Inscribed Circle.

a) Prove that the bisectors of the angles of a triangle all meet at the same point P, and there is a circle centered at P that is tangent to all three sides of the triangle (Fig. 1.46).

b) What is the largest sphere that will pass through a triangular hole with sides 7 in., 8 in., and 9 in. long?

c) (Adapted from [15, page 66 no. 164]). A gardener cut a piece of sod to fill a triangular hole. When he tried to put it in the hole he found that it fit perfectly, but only with the wrong side up. How can he cut the triangle into three pieces so that the shape of each piece is not changed when he turns it over?

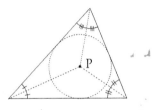

FIGURE 1.46. The Inscribed Circle.

Exercise 1.9.8 A globe is supported by a triangular stand set on a table (see Figure 1.47 below). Pads are placed on the sides of the stand to prevent the globe from getting scratched. The stand is 3 in. tall; its sides form a $30° - 60° - 90°$ triangle whose shortest side is 1 ft. long. Exactly where on the triangle should each pad be placed? If the globe is 2 ft. in diameter, how high is the top of the globe above the table? (See Fig. 1.47. Neglect the thickness of the pads).

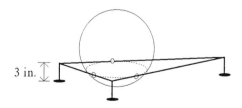

FIGURE 1.47. Globe on a stand.

Exercise 1.9.9 The Circumscribed Circle.

a) Prove that the perpendicular bisectors of the sides of any triangle all meet at a single point P, and there is a circle centered at P that passes through all three vertices of the triangle (Fig. 1.48).

b) State and prove a corresponding result for tetrahedrons and spheres in space.

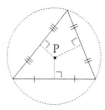

FIGURE 1.48. The Circumscribed Circle.

Exercise 1.9.10 A thin iron triangle with angles 45°, 60°, and 75° is accidentally dropped into a hemispherical tank (fig. 1.49). The tank is ten feet deep and filled with water.

It is proposed to fish the triangle out of the tank by lowering a powerful magnet into the tank with a rope, allowing the magnet to attach itself to the triangle, and then pulling it up. To enable the magnet to reach the triangle the rope must be long enough to reach the triangle from the surface of the water.

What is the minimum length of rope that is required to ensure that one can reach the triangle if the shortest side of the triangle is ten feet long?

FIGURE 1.49. Retrieving a triangle.

Exercise 1.9.11 The Extended Law of Sines.

Given a triangle with sides of length a,b,c opposite vertices A,B,C respectively, prove that

$$\frac{a}{\sin \angle A} = \frac{b}{\sin \angle B} = \frac{c}{\sin \angle C} = d$$

where d is the diameter of the circle that circumscribes the triangle. (Hint: to prove that $a/\sin \angle A = d$, slide A along the circumscribed circle until A and B are opposite ends of a diameter. By Proposition 1.9.4 this does not change $\angle A$ or a.)

Exercise 1.9.12 The Orthocenter.

An *altitude* of a triangle is a line that is perpendicular to one side of the triangle and extends to the opposite vertex.

a) Prove that the three altitudes of any triangle all meet at a single point (Fig. 1.50). The point where the altitudes meet is called the *orthocenter* of the triangle.

b) Let $\triangle ABC$ be a triangle in $\mathbf{E}^3$. Let S_1 be the sphere with diameter $\overline{AB}$, S_2 the sphere with diameter $\overline{AC}$, and S_3 the sphere with diameter $\overline{BC}$. Prove that the intersection $S_1 \cap S_2 \cap S_3$ contains exactly two points P and Q. Show that the plane containing $\triangle ABC$ is the perpendicular bisector of $\overline{PQ}$ and the midpoint of $\overline{PQ}$ is the orthocenter of $\triangle ABC$. (Hint: Show that the perpendicular bisector of $\overline{AB}$ is a plane that contains the circle $S_2 \cap S_3$.)

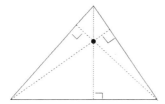

FIGURE 1.50. The Orthocenter.

Exercise 1.9.13 A tripod has three legs $\overline{AB}$, $\overline{AC}$, and $\overline{AD}$, joined at the point A. $AB = 8$in., $AC = 9$in., $AD = 10$in., $\angle CAD = 40^\circ$, $\angle DAB = 50^\circ$, and $\angle BAC = 60^\circ$. How high is the tripod? Will it stand up or will it fall over?

(Hint: The argument in the next two paragraphs leads to a solution of this problem. The student should fill in the details of the argument, supplying any necessary diagrams, and then use the result to calculate the height of the tetrahedron and its center of gravity.

Let $P \in \overline{BCD}$ be the point such that the line $\overleftrightarrow{PA}$ is perpendicular to the plane $\overline{BCD}$. PA is the altitude of the tetrahedron. To locate P, erect triangles $\triangle A_1 BC \cong \triangle ABC$, $\triangle A_2 BD \cong \triangle ABD$, and $\triangle A_3 CD \cong \triangle ACD$ in the plane $\overline{BCD}$, on the outside of $\triangle BCD$. Each of the three lines $\overleftrightarrow{PA_1}$, $\overleftrightarrow{PA_2}$, $\overleftrightarrow{PA_3}$ is an altitude of one of the triangles $\triangle A_1 BC$, $\triangle A_2 BD$, or $\triangle A_3 CD$, so one can locate P by intersecting these three altitudes.

To see why $\overleftrightarrow{PA_1}$ is an altitude of $\triangle A_1 BC$, observe that $\triangle A_1 BC$ can be obtained by revolving $\triangle ABC$ around $\overleftrightarrow{BC}$ until it falls into the plane $\overline{BCD}$.

As the point A revolves, it sweeps out an arc of a circle in a plane that is perpendicular to $\overleftrightarrow{BC}$. This plane contains P (why?), and the intersection of this plane with the plane $\overline{BCD}$ is an altitude of $\triangle A_1 BC$. Therefore the altitude contains P.)

Exercise 1.9.14 Give a definition for the angle between two planes. (Hint: consider vectors that are normal to the planes.) Let H be the plane that contains the points $(1,0,0),(0,1,0)$, and $(0,0,1)$. Use your definition to compute the angle between H and the x,y plane. (Answer: about $55°$).

Give a definition of the angle between a line and a plane and use it to compute the angle between H and the z axis.

1.10   The General Isometry

Theorem 1.10.1 *Every isometry* $f : \mathbf{E}^n \to \mathbf{E}^n$ *is a composition of rotations, translations, and reflections.*

Sketch of Proof. We will sketch the proof for the case $n = 2$; the general case may be treated in much the same way.

Set up a system of coordinates on $\mathbf{E}^2$ in the usual way. Choose a pair of perpendicular lines for coordinate axes, label the four "quadrants", then assign coordinates (x, y) to each point in $\mathbf{E}^2$ by measuring the distance $(\pm y)$ from the point to one of the axes (the "x axis") and the distance $(\pm x)$ from the point to the other axis (the "y axis"). The plus or minus signs depend on which quadrant it is that contains the point.

The "positive end" of the x axis is the set of points with positive x coordinates on the x axis. The "positive end" of the y axis is defined in a similar way.

Let

$$
\begin{aligned}
X &= \text{the x axis,} \\
Y &= \text{the y axis.}
\end{aligned}
$$

An isometry $f : \mathbf{E}^2 \to \mathbf{E}^2$ maps X and Y to another pair of perpendicular lines. Let

$$
\begin{aligned}
X' &= f(X), \\
Y' &= f(Y).
\end{aligned}
$$

We shall regard X' and Y' as a new pair of coordinate axes, labeling the quadrants so that f maps the i-th quadrant in the X, Y coordinate system onto the i−th quadrant in the X', Y' coordinate system for each $i = 1, \ldots, 4$.

Since isometries preserve distance, the distance from X to any point $P \in \mathbf{E}^2$ equals the distance from X' to $f(P)$. Similarly, the distance Y

to P equals the distance from Y' to $f(P)$. It follows that *if a given point* $P \in \mathbf{E}^2$ *has coordinates* (x, y) *in the* X, Y *coordinate system then the point* $f(P)$ *has the same coordinates (x,y) in the* X', Y' *coordinate system!*

Consequently one can compute $f(P)$ for every point $P \in \mathbf{E}^2$ as soon as one knows what f does to the coordinate axes. In particular, *two isometries are the same if they both have the same effect on the coordinate axes* (fig. 1.51).

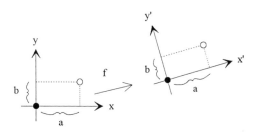

FIGURE 1.51.

Let T be the translation such that

$$T(0, 0) = f(0, 0).$$

T maps X to a line through $f(0, 0)$. Choose a rotation around $f(0, 0)$ that maps $T(X)$ to X'. There are two such rotations, each differing from the other by 180°; let R be the one that takes points on the positive end of X to points on the positive end of X'. The composition $R \circ T$ maps X to X'. Since $(R \circ T)(Y))$ preserves angles it must also map Y to Y'.

If $R \circ T$ maps points on the positive end of Y to points on the positive end of Y' then $R \circ T$ has the same effect on X and Y as f does, which proves that $f = R \circ T$. Otherwise $R \circ T$ maps points on the positive end of Y to points on the negative end of Y'. In this case let F be reflection in X'. $(F \circ R \circ T)$ maps the positive end of X to the positive end of X' and the positive end of Y to the positive end of Y", so $f = F \circ R \circ T$. In either case f is a composition of translations, rotations, and reflections, which completes the proof.

Corollary 1.10.1 *Every isometry is a composition of reflections.*

Proof. By Proposition 1.5.1 and Exercise 1.6.1 every rotation and every translation in $\mathbf{E}^2$ is a composition of reflections. The same facts hold in $\mathbf{E}^n$ (with similar proofs).

Exercise 1.10.1 Show that if $f : \mathbf{E}^2 \to \mathbf{E}^2$ is an orientation-preserving isometry then f is a translation or a rotation.

(Hint: let P be a point such that $f(P) \neq P$. Set up a coordinate system whose origin is at P and whose positive x axis passes through $f(P)$. Let C be the point where the perpendicular bisectors of the line segments $\overline{Pf(P)}$

and $\overline{f(P)f(f(P))}$ intersect. Show that $f(C) = C$, and deduce that f is a rotation about C through an angle equal to $\angle PCf(P)$. If the perpendicular bisectors $\overline{Pf(P)}$ and $\overline{f(P)f(f(P))}$ do not intersect show that f is a translation or a rotation through a $180°$ angle.)

Exercise 1.10.2 Show that every isometry on $\mathbf{E}^2$ or $\mathbf{E}^3$ has an inverse. (Hint: use Theorem 1.10.1 together with Exercise 1.4.1, Equation 1.1, and Equation 1.2 to write down an inverse for an arbitrary isometry).

Corollary 1.10.2 A formula for the general isometry on $\mathbf{R}^2$. *Every isometry $f : \mathbf{R}^2 \to \mathbf{R}^2$ has the form*

$$f(x,y) = (a_0 + x\cos\theta \mp y\sin\theta, b_0 + x\sin\theta \pm y\cos\theta). \qquad (1.6)$$

Here, a_0, b_0, and θ are constants, $f(0,0) = (a_0, b_0)$ in the standard coordinates on $\mathbf{R}^2$, and the plus or minus sign is 'plus' if f preserves orientation or 'minus' if f reverses orientation.

Proof. Let $(a_0, b_0) = f(0,0)$. Let

$$\begin{aligned}
\overrightarrow{v} &= f(1,0) - f(0,0), \\
\overrightarrow{w} &= f(0,1) - f(0,0)
\end{aligned}$$

be the vectors that point from $f(0,0)$ to $f(1,0)$ and $f(0,1)$, respectively. $\overrightarrow{v}$ and $\overrightarrow{w}$ are perpendicular unit vectors since f is an isometry, so there exists an angle θ such that

$$\begin{aligned}
\overrightarrow{v} &= (\cos\theta, \sin\theta) \text{ and} \qquad\qquad (1.7) \\
\overrightarrow{w} &= \pm(-\sin\theta, \cos\theta).
\end{aligned}$$

The sign in the formula for $\overrightarrow{w}$ is 'plus' if f preserves orientation, 'minus' otherwise.

f maps the x axis to an axis X' with origin at (a_0, b_0) and positive end pointing in in the $\overrightarrow{v}$ direction, and it maps the y axis to an axis Y' with origin at (a_0, b_0) and positive end pointing in the $\overrightarrow{w}$ direction. Since each point $f(x, y)$ has the same coordinates in the X', Y' coordinate system as (x, y) has in the x,y coordinate system, it follows that

$$f(x,y) = (a_0, b_0) + x\overrightarrow{v} + y\overrightarrow{w}$$

for each point (x,y) in the x,y plane. Plug Equations 1.7 into this formula; the result is Equation 1.6.

This completes the proof.

1.11   Appendix: The Planimeter

A *planimeter* (see [14, page11]) is a mechanical device for measuring areas of figures in the plane. In its simplest form (Fig. 1.52) it consists of two rods $\overline{XY}$ and $\overline{YZ}$ which are joined together by a hinge at Y and connected to a pivot which is attached to the plane at X. At Z is a pointer which is free to move about on the plane, and there is a wheel W mounted on the rod $\overline{YZ}$ in such a way that it can rotate around $\overline{YZ}$ but it cannot slide along the rod toward either end. The edge of the wheel rests on the plane.

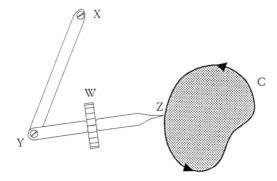

FIGURE 1.52. A Planimeter.

Let C be a simple closed curve[4] in the plane. To measure the area inside C one traces once around the curve with the pointer. As the pointer traces out C the vectors $\overrightarrow{XY}$ and $\overrightarrow{YZ}$ rotate and translate about in the plane. For the planimeter to work correctly it is essential that *in the course of their motion neither $\overrightarrow{XY}$ nor $\overrightarrow{YZ}$ ever rotates a full $360°$ from its original direction.* This condition is easy to achieve in practice if one uses a big enough planimeter and places X sufficiently far away from the curve.

As the pointer moves it drags the wheel along with it, causing the wheel to slide and roll along on the plane. Ignore the sliding, but keep track of how far the wheel rolls by counting the number of revolutions it makes around its axis. (Counterclockwise revolutions count as positive revolutions, clockwise revolutions count as negative). The area inside the curve is obtained by plugging the total number of revolutions into the following formula:

$$\text{area inside } C \;=\; L(2\pi\rho N) \tag{1.8}$$
$$=\; LD$$

where

$$N \;=\; \text{number of revolutions made by the wheel}$$

---

[4]*Simple* means the curve does not cross itself.

$$L = \text{length of } \overline{YZ}$$
$$\rho = \text{radius of the wheel}$$
$$D = \text{distance the wheel rolled.}$$

(N need not be an integer).

To see why Equation 1.8 is true, consider the area that is swept out by the rod $\overline{YZ}$ as it makes an infinitesimal motion in the plane. The infinitesimal motion can be broken down into an infinitesimal translation T and an infinitesimal rotation R (see Fig. 1.53). During the translation the rod sweeps out the interior of the parallelogram bounded by $\overline{YZ}$ and $T(\overline{YZ})$. During the rotation the rod sweeps out the interior of the circular sector between $T(\overline{YZ})$ and $R(T(\overline{YZ}))$. The total infinitesimal area that it sweeps out is the sum of these areas,

$$\Delta(\text{area}) = (\text{area of the parallelogram}) + (\text{area of the sector})$$
$$= L\Delta h + \frac{L^2 \Delta \theta}{2} \tag{1.9}$$

where

$$\Delta h = (\text{height of the parallelogram}),$$

and

$$\Delta \theta = (\text{angle of the rotation}).$$

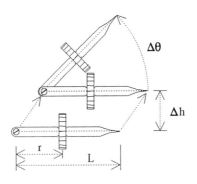

FIGURE 1.53.

The wheel rolls on the plane in directions perpendicular to $\overline{YZ}$ and slides on the plane in directions parallel to $\overline{YZ}$. The distance it rolls during the translation plus the distance it rolls during the rotation is:

$$\Delta D = \Delta h + r\Delta \theta. \tag{1.10}$$

where

$$r = \text{distance from the hinge } Y \text{ to the wheel.}$$

Thus by equations 1.9 and 1.10, $\overline{YZ}$ sweeps out the total infinitesimal area $\Delta(\text{area})$ that is given by the formula,

$$\begin{aligned}
\Delta(\text{area}) &= L\Delta D - Lr\Delta\theta + \frac{L^2\Delta\theta}{2} \\
&= L\Delta D + L\left(\frac{L}{2} - r\right)\Delta\theta.
\end{aligned}$$

Summing all these infinitesimal areas we find the total area swept out by the rod as the pointer moves around the curve.

$$\begin{pmatrix}\text{total area swept} \\ \text{out by } \overline{YZ}\end{pmatrix} = LD + L\left(\frac{L}{2} - r\right)\theta \qquad (1.11)$$

where

$$\begin{aligned}
D &= (\text{total distance rolled by the wheel}), \text{ and} \\
\theta &= \left(\text{total angle of rotation of the vector } \overrightarrow{YZ}.\right)
\end{aligned}$$

Now the curve is closed, so the rods must return to their starting position when the pointer reaches the end of the curve. Therefore θ must be an integral multiple of $360°$. $\overrightarrow{YZ}$ is not allowed to rotate through an entire $360°$ angle, so it follows that the total angle θ must be zero,

$$\begin{aligned}
\begin{pmatrix}\text{total area swept} \\ \text{out by } \overline{YZ}\end{pmatrix} &= LD \\
&= L(2\pi\rho N)
\end{aligned}$$

by Equation 1.11. (The second line follows from the obvious fact that the distance, D, that the wheel rolls is the product of its circumference, $2\pi\rho$, and the number N of revolutions that it makes.)

To prove Equation 1.8 it remains to show that the area swept out by $\overline{YZ}$ is the area inside C.

As the wheel rolls distances contributed by counterclockwise revolutions count positively while distance distances contributed by clockwise revolutions count negatively. The same goes for area: positive revolutions contribute positive area and negative revolutions contribute negative area. The net effect is shown in Figure 1.54: the area swept out as the rod moves from Y_1Z_1 to Y_2Z_2 is positive while the area swept out as it moves back to Y_1Z_1 is negative. Thus the area inside C is positive, the area inside the curve C' traced out by Y is negative, and the area between the two curves cancels out,

$$\begin{pmatrix}\text{total area swept} \\ \text{out by } \overline{YZ}\end{pmatrix} = (\text{area inside } C) - (\text{area inside } C').$$

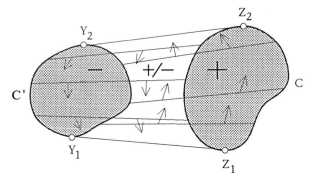

FIGURE 1.54.

C' is an arc of a circle centered at X since the vector $\overrightarrow{XY}$ is fixed to the plane at X. But $\overrightarrow{XY}$ is not allowed to rotate through full $360°$ angle; it follows that the area inside C' is zero. Hence

$$\begin{pmatrix} \text{total area swept} \\ \text{out by } \overline{YZ} \end{pmatrix} = (\text{area inside } C).$$

This completes the proof of Equation 1.8.

2
Spherical Geometry

2.1 Geodesics

In any geometrical setting where it makes sense to talk about the distance between points, the most important curves are the geodesics.

Definition 2.1.1 Geodesics. The shortest curve connecting two points in a space is a *geodesic* in that space.

Example 2.1.1 A geodesic connecting two points on the globe can be found by stretching a piece of string across the globe between the points and pulling it tight. The geodesic connecting Los Angeles to London passes northeast through central Canada, turns east across the southern tip of Greenland, and arrives in London heading southeast. (Try it!) Ships and airliners save fuel by following such "great circle routes" when traveling long distances (fig. 2.1).

It should be emphasized that *one looks only at curves that lie entirely in the space* when searching for geodesics in a space S. The fact that there may be shorter curves outside of S is irrelevant – one treats S as if it were the entire universe. For instance one could find a shorter path from Los Angeles to London than the one in Example 2.1.1 by burrowing through the earth, but that does not matter since such a path would take one out of the "universe" which, in this case, is the surface of the globe.

There may be more than one geodesic connecting a given pair of points. For example, there are infinitely many geodesics connecting the north and south poles on the globe.

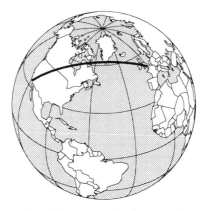

FIGURE 2.1. Geodesic connecting Los Angeles to London.

One can connect shorter geodesics together to get longer ones in the same way that one joins line segments to get a line. In general we say that a curve is a "geodesic" if every sufficiently small segment of the curve has minimum length in the sense of Definition 2.1.1. Such geodesics minimize length over short distances but may fail to minimize length over long distances.

Example 2.1.2 If you roll up a piece of paper into a cylinder, connect two points on the cylinder with a piece of string, and pull the string tight, you will get a geodesic connecting the two points. There are infinitely many ways to do this, depending on the number of times the string winds around the cylinder, but only one (or, in certain cases, two) of these geodesics minimizes the length between the two points.

A geodesic may even intersect itself. You can create an example of this by wrapping a string tightly around a cone (Fig. 2.2).

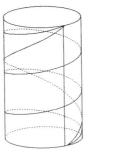

FIGURE 2.2. Geodesics.

Definition 2.1.2 Geodesic Triangle. A *geodesic triangle* is a "triangle" which consists of three vertices connected by geodesics.

A little experimentation shows that the angles of a geodesic triangle on a curved surface need not add up to 180°.

Example 2.1.3 Make a paper cone by joining the edges of a circular sector (Fig. 2.3). Mark three points A, B, and C on the cone, and join them with geodesic segments by flattening the cone out on a table and connecting the points with line segments. These line segments remain shortest curves on the paper even when it is lifted off the table, unflattened, and bent into a cone, because flattening or unflattening the paper does not stretch or shrink it, and so does not distort lengths within the paper. Thus $\triangle ABC$ is a geodesic triangle on the cone.

Let θ be the angle subtended by the circular sector. If the vertex of the cone lies in the interior of $\triangle ABC$ you will find that

$$\angle A + \angle B + \angle C = 540° - \theta. \tag{2.1}$$

In particular $\angle A + \angle B + \angle C > 180°$ if $\theta < 360°$.

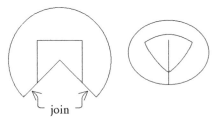

FIGURE 2.3. Triangle on a cone.

You can make a circular sector that subtends an angle $\theta > 360°$ by glueing together two smaller sectors. If you join together the edges of this sector you will get a saddle-shaped surface (Fig. 2.4). Equation 2.1 applies to this case as well, so the angles of a geodesic triangle on the saddle-shaped surface add up to less than 180° if the vertex of the sector is in the triangle's interior.

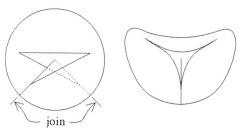

FIGURE 2.4. Triangle on a saddle.

Based on these examples one might guess that the angles of a geodesic triangle on a surface add up to less than 180° if the surface is saddle-shaped, or more than 180° if the surface is is bowl shaped like a cone. Such a result would enable two-dimensional beings who lived inside the surface to discover whether it is saddle- or bowl- shaped, without ever going outside the surface, by adding up the angles of a geodesic triangle and comparing the sum with 180°. A similar procedure would enable three-dimensional beings such as ourselves to discover whether or not their universe is curved without having to leave their universe to make the measurement. (Our own universe *is* curved, but you have to look at very large triangles to detect the curvature.)

The famous *Gauss-Bonnet Theorem* says that the conjectures in the previous paragraph are true, at least on surfaces that are sufficiently smooth; it is proved in courses in differential geometry.[1] We shall content ourselves with proving it in the special case where the surface is a sphere (Theorem 2.3.1 on page 51).

Exercise 2.1.1 Prove Equation 2.1 for surfaces built from sectors subtending an arbitrary angle $0° < \theta < 540°$. Show that if $\theta \geq 540°$ then no geodesic triangle in the surface has the vertex of the sector in its interior. What is the sum of the angles of a geodesic triangle on these surfaces if the vertex of the surface is not in its interior?

2.2 Geodesics on Spheres

Definition 2.2.1 A *great circle* is the intersection of a sphere and a plane that passes through the center of the sphere. All other circles on the sphere are "small circles" (Fig. 2.5).

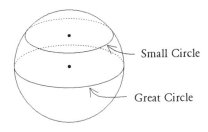

FIGURE 2.5.

Let A and B be two points on the sphere and let C be its center. The arc $\overset{\frown}{AB}$ subtended by the $\angle ACB$ is a segment of a great circle with length

$$\text{length}(\overset{\frown}{AB}) = R(\angle ACB)$$

where R is the radius of the sphere and $\angle ACB$ is measured in radians.

---

[1]See [17, Chap. 7 §8].

It is convenient to use spherical coordinates when computing the lengths of curves in the sphere. Set up a system of rectangular coordinates on $\mathbf{E}^3$ with the origin at the center C of the sphere and the positive z axis piercing the sphere at A. The spherical coordinates of a point P on the sphere are (R, θ, ϕ) where

R is the radius of the sphere,
$\phi = \angle PCA$, and
θ is the angle between the positive x axis and the projection of
$\overrightarrow{CP}$ into the x,y plane.

(See Fig. 2.6).

Elementary trigonometry shows that spherical coordinates are related to rectangular coordinates by the formulas:

$$
\begin{aligned}
x &= R\sin\phi\cos\theta \\
y &= R\sin\phi\sin\theta \\
z &= R\cos\phi.
\end{aligned}
$$

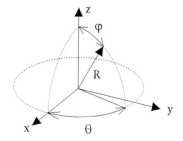

FIGURE 2.6. Spherical Coordinates.

Theorem 2.2.1 *The shortest path between two points on a sphere is an arc of a great circle.*

Proof. Let $\sigma : [a, b] \rightarrow S$ be a parametrized curve in S with

$$\sigma(a) = A \text{ and } \sigma(b) = B.$$

Write $\sigma(t) = (x(t), y(t), z(t))$ in rectangular coordinates. The length of the curve σ is given by the length formula

$$\text{length}(\sigma) = \int_a^b \sqrt{x'(t)^2 + y'(t)^2 + z'(t)^2}\, dt. \qquad (2.2)$$

Changing to spherical coordinates,

$$\begin{aligned}
x(t) &= R\sin\phi(t)\cos\theta(t) \\
y(t) &= R\sin\phi(t)\sin\theta(t) \\
z(t) &= R\cos\phi(t).
\end{aligned} \tag{2.3}$$

Plug the derivatives

$$\begin{aligned}
x'(t) &= R(\cos\phi(t)\cos\theta(t)\phi'(t) - \sin\phi(t)\sin\theta(t)\theta'(t)) \\
y'(t) &= R(\cos\phi(t)\sin\theta(t)\phi'(t) + \sin\phi(t)\cos\theta(t)\theta'(t)) \\
z'(t) &= -R\sin\phi(t)\phi'(t)
\end{aligned}$$

into the integrand in Equation 2.2 and get

$$\begin{aligned}
\text{length}(\sigma) &= \int_a^b R\sqrt{\phi'(t)^2 + \sin^2\phi(t)\theta'(t)^2}\,dt \\
&\geq \int_a^b R\phi'(t)\,dt \\
&= R(\phi(b) - \phi(a)) \\
&= R(\angle BCA) \\
&= \text{length}(\overset{\frown}{AB}).
\end{aligned}$$

with strict inequality unless $\theta'(t) = 0$ or $\sin^2\phi(t) = 0$ for all t, that is, unless σ never leaves the arc $\overset{\frown}{AB}$.

This completes the proof.

The geometry of geodesics on the sphere is different from the geometry of lines in the plane. For instance a sphere has no "parallel" geodesics since two great circles always intersect at a pair of diametrically opposite points (Fig. 2.7).

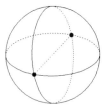

FIGURE 2.7. Great circles always intersect.

2.3   The Six Angles of a Spherical Triangle

A *spherical triangle* is a geodesic triangle on the surface of a sphere. Let $\triangle ABC$ be a spherical triangle with side a opposite vertex A, side b opposite vertex B, and side c opposite vertex C, on a sphere with center at O. $\triangle ABC$ has six angles: three *arc angles* $\angle a$, $\angle b$, and $\angle c$ and three *vertex angles* $\angle A$, $\angle B$, $\angle C$ (Fig. 2.8).

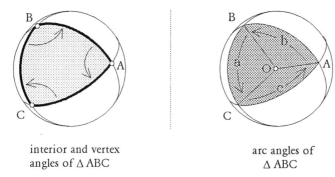

interior and vertex
angles of $\triangle ABC$

arc angles of
$\triangle ABC$

FIGURE 2.8. Six angles.

Arc angles measure the angles subtended by the sides of the triangle. $\angle a = \angle(\overrightarrow{OB}, \overrightarrow{OC})$, $\angle b = \angle(\overrightarrow{OC}, \overrightarrow{OA})$, and $\angle c = \angle(\overrightarrow{OA}, \overrightarrow{OB})$.[2]

Vertex angles measure three things at once. $\angle A$ equals

1. the angle between the arcs $\overset{\frown}{AB}$ and $\overset{\frown}{AC}$ at A,

2. the angle between a vector $\overrightarrow{V}$ that is tangent to $\overset{\frown}{AB}$ at A, and a vector $\overrightarrow{W}$ that is tangent to $\overset{\frown}{AC}$ at A,

3. the angle between the planes $\overline{ABO}$ and $\overline{ACO}$.

The first and second items in the above list are equal by definition. The second and third items are equal because $\overrightarrow{V}$ and $\overrightarrow{W}$ are perpendicular to the line $\overleftrightarrow{AO}$ where the planes $\overline{ABO}$ and $\overline{ACO}$ intersect (Fig. 2.9).

Similar statements hold for $\angle B$ and $\angle C$.

To simplify the notation from now on, set

$$\overrightarrow{A} = \overrightarrow{OA}, \ \overrightarrow{B} = \overrightarrow{OB}, \text{ and } \overrightarrow{C} = \overrightarrow{OC}.$$

We shall regard $\angle(\overrightarrow{A}, \overrightarrow{B})$ and $\angle(\overrightarrow{B}, \overrightarrow{A})$ as representing the same angle; for our purposes angles of a spherical triangle have no orientation.

Lemma 2.3.1 *In a spherical triangle* $\triangle ABC$,

---

[2] $\angle(\overrightarrow{OA}, \overrightarrow{OB})$ denotes the angle between the vectors $\overrightarrow{OA}$ and $\overrightarrow{OB}$, etc. All spherical angles measure between $0°$ and $180°$.

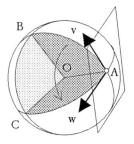

vertex angle $\angle A$

FIGURE 2.9. Vertex angle.

1. the arc angles are

$$
\begin{aligned}
\angle a &= \angle(\overrightarrow{B}, \overrightarrow{C}) \\
\angle b &= \angle(\overrightarrow{C}, \overrightarrow{A}) \\
\angle c &= \angle(\overrightarrow{A}, \overrightarrow{B}).
\end{aligned}
$$

2. the vertex angles are

$$
\begin{aligned}
\angle A &= \angle(\overrightarrow{A} \times \overrightarrow{B}, \overrightarrow{A} \times \overrightarrow{C}) \\
\angle B &= \angle(\overrightarrow{B} \times \overrightarrow{C}, \overrightarrow{B} \times \overrightarrow{A}) \\
\angle C &= \angle(\overrightarrow{C} \times \overrightarrow{A}, \overrightarrow{C} \times \overrightarrow{B}).
\end{aligned}
$$

where $\times$ is the "cross product".

Proof. 1. This is just a restatement of the definition of arc angles.

2.: $\overrightarrow{A} \times \overrightarrow{B}$ is perpendicular to $\overline{ABO}$, and $\overrightarrow{A} \times \overrightarrow{C}$ is perpendicular to $\overline{ACO}$. This *almost* proves that $\angle A = \angle(\overrightarrow{A} \times \overrightarrow{B}, \overrightarrow{A} \times \overrightarrow{C})$ since an angle between two planes equals the angle between their normal vectors. The problem is that two intersecting planes actually determine two supplementary angles,[3] so we need to make sure that $\angle(\overrightarrow{A} \times \overrightarrow{B}, \overrightarrow{A} \times \overrightarrow{C})$ equals $\angle A$ and not $180° - \angle A$.

Let $\overrightarrow{V}$ be tangent to $\overset{\frown}{AB}$ and let $\overrightarrow{W}$ be tangent to $\overset{\frown}{AC}$ at A.[4]

$$
\angle A = \angle(\overrightarrow{V}, \overrightarrow{W}). \tag{2.4}
$$

---

[3]Supplementary angles add up to $180°$.

[4]More precisely, $\overrightarrow{V}$ is the initial velocity of a particle traveling from A to B along $\overset{\frown}{AB}$. A similar definition applies to W.

Since $\overrightarrow{V}$ and $\overrightarrow{W}$ are perpendicular to $\overrightarrow{A}$ the right hand rule says that $\overrightarrow{A} \times \overrightarrow{V}$ points in the direction obtained by rotating $\overrightarrow{V}$ ninety degrees to the right around $\overrightarrow{A}$. Similarly $\overrightarrow{A} \times \overrightarrow{W}$ points $90°$ to the right of $\overrightarrow{W}$ around $\overrightarrow{A}$. Rotations do not change angles, so

$$\angle(\overrightarrow{V}, \overrightarrow{W}) = \angle(\overrightarrow{A} \times \overrightarrow{V}, \overrightarrow{A} \times \overrightarrow{W}). \tag{2.5}$$

But $\overrightarrow{A} \times \overrightarrow{V}$ points the same direction[5] as $\overrightarrow{A} \times \overrightarrow{B}$, and $\overrightarrow{A} \times \overrightarrow{W}$ points the same direction as $\overrightarrow{A} \times \overrightarrow{C}$. Thus

$$\angle(\overrightarrow{A} \times \overrightarrow{V}, \overrightarrow{A} \times \overrightarrow{W}) = \angle(\overrightarrow{A} \times \overrightarrow{B}, \overrightarrow{A} \times \overrightarrow{C}). \tag{2.6}$$

Equations 2.4, 2.5, 2.6 imply that

$$\angle A = \angle(\overrightarrow{A} \times \overrightarrow{B}, \overrightarrow{A} \times \overrightarrow{C}).$$

Similar arguments show that $\angle B = \angle(\overrightarrow{B} \times \overrightarrow{C}, \overrightarrow{B} \times \overrightarrow{A})$ and $\angle C = \angle(\overrightarrow{C} \times \overrightarrow{A}, \overrightarrow{C} \times \overrightarrow{B})$.
This completes the proof.

Theorem 2.3.1 *In a spherical triangle $\triangle ABC$*

$$\angle A + \angle B + \angle C = \pi + \frac{area(\triangle ABC)}{R^2} \tag{2.7}$$

in radians, where R is the radius of the sphere.

In particular $\angle A + \angle B + \angle C > 180°$.
Proof. Given a point P on the sphere, let $-P$ be the point on the opposite end of a diameter from P. Two great circles meeting at an angle θ at P bound a sector of the sphere with vertices P and $-P$. Fig. 2.10 shows that the ratio of the sector's area to the sphere's area is the same as the ratio of the angle θ to an angle subtending a full circle:

$$\frac{area(sector)}{area(sphere)} = \frac{\theta}{2\pi}.$$

Since the area of the sphere is $4\pi R^2$ it follows that

$$area(sector) = 2R^2\theta.$$

---

[5]Since $\overrightarrow{A}$, $\overrightarrow{B}$, and $\overrightarrow{B}$ lie in the same plane; use the right-hand rule.

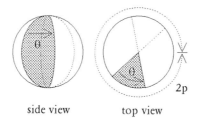

FIGURE 2.10. Area of sector $= 2 \times$ (angle of sector).

Each vertex angle of $\triangle ABC$ subtends a sector on the sphere; call them the "A-sector", the "B-sector", and the "C-sector". Let H_C be the hemisphere that contains C and is bounded by $\widehat{AB}$. H_C contains both the A- and the B- sectors but only part of the C-sector. Fig. 2.11 shows two views of the three sectors. In the view on the right the sphere has been rotated so that H_C is the visible hemisphere; part of the C-sector wraps around to the back of the sphere behind the visible hemisphere.

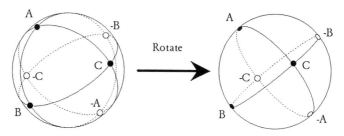

FIGURE 2.11. Rotating the sphere.

From Fig. 2.11 one can see that

$$
\begin{aligned}
\text{area}(H_C) &= \text{area}(\triangle ABC) + \text{area}(\triangle(-A)BC) && (2.8)\\
&\quad + \text{area}(\triangle A(-B)C) + \text{area}(\triangle(-A)(-B)C) \\
\text{area}(\text{A-sector}) &= \text{area}(\triangle ABC) + \text{area}(\triangle(-A)BC), && (2.9)\\
\text{area}(\text{B-sector}) &= \text{area}(\triangle ABC) + \text{area}(\triangle A(-B)C), && (2.10)\\
\text{area}(\text{C-sector}) &= \text{area}(\triangle ABC) + \text{area}(\triangle AB(-C)). && (2.11)
\end{aligned}
$$

Reflection through the center of the sphere is an isometry f that takes each point P on the sphere to the point $-P$ at the opposite end of a diameter.[6] $f(\triangle AB(-C)) = \triangle(-A)(-B)C$ so, since f is an isometry, it follows that

$$
\text{area}(\triangle AB(-C)) = \text{area}(\triangle(-A)(-B)C).
$$

---

[6]If the sphere is centered at the origin then $f(x,y,z) = (-x,-y,-z)$, in coordinates.

Hence, by Equation 2.11,

$$\text{area(C-sector)} = \text{area}(\triangle ABC) + \text{area}(\triangle(-A)(-B)C). \qquad (2.12)$$

Comparing the sum of Equations 2.9, 2.10, and 2.12 with Equation 2.8, we find that

$$\text{area(A-sector)} + \text{area(B-sector)} + \text{area(C-sector)} \qquad (2.13)$$
$$= \text{area}(H_C) + 2\text{area}(\triangle ABC).$$

(see Fig. 2.12). Since

$$\begin{aligned}
\text{area(A-sector)} &= 2R^2 \angle A \\
\text{area(B-sector)} &= 2R^2 \angle B \\
\text{area(C-sector)} &= 2R^2 \angle C
\end{aligned}$$

and

$$\text{area}(H_C) = 2\pi R^2 = (1/2)\text{area(sphere)},$$

Equation 2.13 says that

$$2R^2 \angle A + 2R^2 \angle B + 2R^2 \angle C = 2\pi R^2 + 2\text{area}(\triangle ABC).$$

Divide by $2R^2$ to get Equation 2.7.
    This completes the proof.

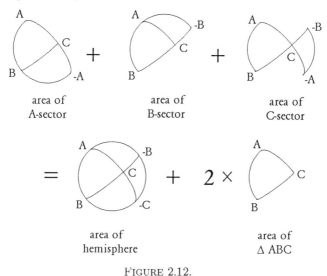

FIGURE 2.12.

Corollary 2.3.1 It is impossible to have an isometry between a region on a sphere and a region in the plane, so long as the region in the sphere includes at least one geodesic triangle together with its interior.

Proof. Isometries preserve distances so they map shortest curves to shortest curves. Therefore such an isometry would map geodesic triangles on the sphere to geodesic triangles on the plane. But isometries also preserve angles.[7] Therefore the sum of the angles of the spherical triangle would have to equal the sum of the angles of a planar triangle, and this contradicts Theorem 2.3.1.

2.4   The Law of Cosines for Sides

Let $\triangle ABC$ be a spherical triangle with side a opposite vertex A, side b opposite vertex B, and side c opposite vertex C.

Proposition 2.4.1 The Law of Cosines for Sides.

$$\cos \angle a = \cos \angle b \cos \angle c + \sin \angle b \sin \angle c \cos \angle A.$$

Proof. We use the dot product to compute $\cos(\angle A)$. By Lemma 2.3.1, $\angle A = \angle(\overrightarrow{A} \times \overrightarrow{B}, \overrightarrow{A} \times \overrightarrow{C})$, so:

$$(\overrightarrow{A} \times \overrightarrow{B}) \cdot (\overrightarrow{A} \times \overrightarrow{C}) = |\overrightarrow{A} \times \overrightarrow{B}||\overrightarrow{A} \times \overrightarrow{C}| \cos \angle A.$$

Also $\angle(\overrightarrow{A}, \overrightarrow{B}) = \angle c$ and $\angle(\overrightarrow{A}, \overrightarrow{C}) = \angle b$ so

$$|\overrightarrow{A} \times \overrightarrow{B}| = |\overrightarrow{A}||\overrightarrow{B}| \sin \angle c,$$
$$|\overrightarrow{A} \times \overrightarrow{C}| = |\overrightarrow{A}||\overrightarrow{C}| \sin \angle b.$$

Thus

$$\begin{aligned}
(\overrightarrow{A} \times \overrightarrow{B}) \cdot (\overrightarrow{A} \times \overrightarrow{C}) &= (|\overrightarrow{A}||\overrightarrow{B}| \sin \angle c)(|\overrightarrow{A}||\overrightarrow{C}| \sin \angle b) \cos \angle A \\
&= R^4 \sin \angle b \sin \angle c \cos \angle A
\end{aligned}$$

$$(2.14)$$

since $|\overrightarrow{A}| = |\overrightarrow{B}| = |\overrightarrow{C}| = R$. A standard identity[8] from vector algebra enables us to simplify the left-hand side of Equation 2.14:

$$(\overrightarrow{A} \times \overrightarrow{B}) \cdot (\overrightarrow{A} \times \overrightarrow{C}) = \det \begin{pmatrix} \overrightarrow{A} \cdot \overrightarrow{A} & \overrightarrow{A} \cdot \overrightarrow{C} \\ \overrightarrow{B} \cdot \overrightarrow{A} & \overrightarrow{B} \cdot \overrightarrow{C} \end{pmatrix}$$

---

[7]We have proved this only in the plane but it is true in general. The reason is that any angle can be subtended by arcs of infinitesimally small length. Such arcs are basically line segments, so the angle between them can be computed from infinitesimal lengths by using the law of cosines. Since isometries preserve infinitesimal lengths they must, therefore, preserve angles.

[8]$(\overrightarrow{A} \times \overrightarrow{B}) \cdot (\overrightarrow{C} \times \overrightarrow{D}) = (\overrightarrow{A} \cdot \overrightarrow{C})(\overrightarrow{B} \cdot \overrightarrow{D}) - (\overrightarrow{A} \cdot \overrightarrow{D})(\overrightarrow{B} \cdot \overrightarrow{C}).$

$$= \det \begin{pmatrix} |\vec{A}|^2 & |\vec{A}||\vec{C}|\cos \angle b \\ |\vec{B}||\vec{A}|\cos \angle c & |\vec{B}||\vec{C}|\cos \angle a \end{pmatrix}$$

$$= \det \begin{pmatrix} R^2 & R^2 \cos \angle b \\ R^2 \cos \angle c & R^2 \cos \angle a \end{pmatrix}$$

$$= R^4 \cos \angle a - R^4 \cos \angle b \cos \angle c.$$

Combining this result with Equation 2.14 one has

$$R^4 \cos \angle a - R^4 \cos \angle b \cos \angle c = R^4 \sin \angle b \sin \angle c \cos \angle A.$$

Finally divide by R^4, then add $\cos \angle b \cos \angle c$ to both sides to finish the proof.

Remark. When using the law of cosines to solve a triangle the accuracy of the results can easily be ruined by rounding off, so it is best to carry out all calculations to at least five or six decimal places. Better still, use the memories in your calculator to store all your intermediate results, and your answers will be as accurate as your calculator can make them.

Exercise 2.4.1 SAS and SSS.
  Solve the spherical triangle $\triangle ABC$ if
  a)$\angle A = 60°$, $\angle b = 70°$, $\angle c = 80°$.
  b)$\angle a = 60°$, $\angle b = 70°$ and $\angle c = 80°$.

2.5   The Dual Spherical Triangle

The importance of the dual triangle lies in the fact that its vertices correspond to the sides of the original triangle and its sides correspond to the vertices of the original triangle. Every theorem about the sides of the original triangle gives us, for free, a theorem about the vertices of the dual triangle, and every theorem about the vertices of the original triangle gives us a theorem about the sides of the dual triangle. This cuts the amount of work that needs to be done by about half.

"Duality" is used in many parts of mathematics. We will meet it again in Chapter 4.

Definition 2.5.1 Dual Spherical Triangle. Let $\triangle ABC$ be a spherical triangle. The *dual triangle* $^*\triangle\, ABC$ is the spherical triangle $\triangle A^* B^* C^*$ where

$$\vec{A^*} = \frac{\vec{B} \times \vec{C}}{R \sin \angle a},$$

$$\vec{B^*} = \frac{\vec{C} \times \vec{A}}{R \sin \angle b},$$

$$\overrightarrow{C^*} \;=\; \frac{\overrightarrow{A} \times \overrightarrow{B}}{R\sin\angle c}.$$

Here R is the radius of the sphere and, as in the rest of this chapter, $\overrightarrow{A^*} = \overrightarrow{OA^*}$, $\overrightarrow{B^*} = \overrightarrow{OB^*}$ and $\overrightarrow{C^*} = \overrightarrow{OC^*}$ where O is the center of the sphere. The purpose of the scalars $R\sin\angle a$, $R\sin\angle b$, and $R\sin\angle c$ in the definition is to adjust the lengths of the vectors so that $|\overrightarrow{A^*}| = |\overrightarrow{B^*}| = |\overrightarrow{C^*}| = R$. (Recall that $|\overrightarrow{V} \times \overrightarrow{W}| = |\overrightarrow{V}||\overrightarrow{W}|\sin\angle(\overrightarrow{V},\overrightarrow{W})$ for any two vectors $\overrightarrow{V}$ and $\overrightarrow{W}$ and $\angle a = \angle(\overrightarrow{B},\overrightarrow{C})$, $\angle b = \angle(\overrightarrow{C},\overrightarrow{A})$ and $\angle c = \angle(\overrightarrow{A},\overrightarrow{B})$ by Proposition 2.3.1.)

$\overrightarrow{A^*}$ is perpendicular to the plane $\overline{BCO}$ containing side a of the original triangle. Likewise $\overrightarrow{B}$ is perpendicular to the plane containing b and $\overrightarrow{C}$ is perpendicular to the plane containing c. In this sense the vertices of the dual triangle correspond to the sides of the the original triangle. The next proposition shows that in the same sense the sides of the dual triangle correspond to the vertices of the original triangle. Finally Corollary 2.5.3 says that each of the angles on the dual triangle is supplementary[9] to the corresponding angle on the original triangle.

Proposition 2.5.1 *Let* A^*, B^*, C^* *be the vertices of the dual triangle in Definition 2.5.1. Let* a^* *be the side opposite vertex* A^*, b^* *the side opposite vertex* B^*, *and* c^* *the side opposite vertex* C^*. *Then*

$$\overrightarrow{A} \;=\; s\left(\frac{\overrightarrow{B^*} \times \overrightarrow{C^*}}{R\sin\angle a^*}\right)$$

$$\overrightarrow{B} \;=\; s\left(\frac{\overrightarrow{C^*} \times \overrightarrow{A^*}}{R\sin\angle b^*}\right)$$

$$\overrightarrow{C} \;=\; s\left(\frac{\overrightarrow{A^*} \times \overrightarrow{B^*}}{R\sin\angle c^*}\right)$$

where [10]

$$s = \pm1 \text{ is the sign of } \det[\overrightarrow{A},\overrightarrow{B},\overrightarrow{C}].$$

---

[9]Supplementary angles add up to 180°.

[10]$\det[\overrightarrow{A},\overrightarrow{B},\overrightarrow{C}]$ is the determinant of the 3×3 matrix whose rows are the vectors $\overrightarrow{A}$, $\overrightarrow{B}$, and $\overrightarrow{C}$.

Proof.

$$\vec{B^*} \times \vec{C^*} = \left(\frac{\vec{C} \times \vec{A}}{\sin \angle b}\right) \times \left(\frac{\vec{A} \times \vec{B}}{\sin \angle c}\right)$$

$$= \frac{(\vec{C} \times \vec{A}) \times (\vec{A} \times \vec{B})}{\sin \angle b \sin \angle c}. \tag{2.15}$$

Use the the vector identity

$$\vec{X} \times (\vec{Y} \times \vec{Z}) = (\vec{X} \cdot \vec{Z})\vec{Y} - (\vec{X} \cdot \vec{Y})\vec{Z}$$

to multiply out the numerator on the right hand-side of Equation 2.15.

$$(\vec{C} \times \vec{A}) \times (\vec{A} \times \vec{B}) = \left((\vec{C} \times \vec{A}) \cdot \vec{B}\right)\vec{A} - \left((\vec{C} \times \vec{A}) \cdot \vec{A}\right)\vec{B}$$

$$= \vec{A}\det[\vec{A}, \vec{B}, \vec{C}] + 0.$$

Hence

$$\vec{B^*} \times \vec{C^*} = \vec{A}\left(\frac{\det[\vec{A}, \vec{B}, \vec{C}]}{\sin \angle b \sin \angle c}\right),$$

so

$$\vec{A} = (\vec{B^*} \times \vec{C^*})\left(\frac{\sin \angle b \sin \angle c}{\det[\vec{A}, \vec{B}, \vec{C}]}\right).$$

Thus

$$\vec{A} \text{ points in the same direction as } s\left(\frac{\vec{B^*} \times \vec{C^*}}{\sin \angle a^*}\right)$$

$(\sin \angle a^*, \sin \angle b, \sin \angle c > 0$ since $0° < \angle a^*, \angle b, \angle c < 180°)$. Moreover

$$R = |\vec{A}| \text{ and}$$

$$R = \frac{|\vec{B^*} \times \vec{C^*}|}{R \sin \angle a^*}$$

since $|\vec{B^*}| = |\vec{C^*}| = R$ and $\angle(\vec{B^*}, \vec{C^*}) = \angle a^*$. Therefore

$$\vec{A} = s\left(\frac{\vec{B^*} \times \vec{C^*}}{\sin \angle a^*}\right).$$

The other equations follow in a similar way.
This completes the proof.

Corollary 2.5.1 The Dual of the Dual Triangle
  The dual of $^*\triangle ABC$ *is*

$$^*(^*\triangle ABC) = \begin{cases} \triangle ABC & \text{if } \det[\overrightarrow{A}, \overrightarrow{B}, \overrightarrow{C}] \geq 0 \\ \triangle(-A)(-B)(-C) & \text{if } \det[\overrightarrow{A}, \overrightarrow{B}, \overrightarrow{C}] \leq 0 \end{cases}$$

Corollary 2.5.2
$$^*(^*\triangle ABC)) \cong \triangle ABC.$$

Proof. By Corollary 2.5.1 either $^*(^*\triangle ABC))$ equals $\triangle ABC$ or it equals the reflection of $\triangle ABC$ through the center of the sphere.

Corollary 2.5.3 *Let* A^*, B^* *and* C^* *be the vertices of* $^*\triangle ABC$ *as in Definition 2.5.1. Let* a^* *be the side opposite vertex* A^*, b^* *the side opposite vertex* B^*, *and* c^* *the side opposite vertex* C^*. *Then*

$$\begin{aligned} \angle A + \angle a^* &= \angle A^* + \angle a &= 180^o, \\ \angle B + \angle b^* &= \angle B^* + \angle b &= 180^o, \\ \angle C + \angle c^* &= \angle C^* + \angle c &= 180^o. \end{aligned} \qquad (2.16)$$

Proof. By Definition 2.5.1 and Lemma 2.3.1

$$\begin{aligned} \angle a^* &= \angle(\overrightarrow{B^*}, \overrightarrow{C^*}) \\ &= \angle(\overrightarrow{C} \times \overrightarrow{A}, \overrightarrow{A} \times \overrightarrow{B}) \\ &= \angle(-\overrightarrow{A} \times \overrightarrow{C}, \overrightarrow{A} \times \overrightarrow{B}) \\ &= 180^o - \angle(\overrightarrow{A} \times \overrightarrow{C}, \overrightarrow{A} \times \overrightarrow{B}) \\ &= 180^o - \angle A. \end{aligned}$$

Thus
$$\angle A + \angle a^* = 180^o.$$

Applying this result to the dual $\triangle A^{**}B^{**}C^{**}$ of $\triangle A^*B^*C^*$, we have

$$\angle A^* + \angle a^{**} = 180^o, \qquad (2.17)$$

where a^{**} is the side opposite $\angle A^{**}$. But $\angle a^{**} \cong \angle a$, for corollary 2.5.2 says that $\triangle A^{**}B^{**}C^{**}$ is congruent to $\triangle ABC$. Hence

$$\angle A^* + \angle a = 180^o.$$

by Equation 2.17.
  This proves the first Equation 2.16. The other two follow in a similar way.

Exercise 2.5.1 a) Let

$$A = (0,0,1),$$
$$B = (\frac{\sqrt{2}}{2}, 0, \frac{\sqrt{2}}{2}),$$
$$C = (0, \frac{\sqrt{3}}{2}, \frac{1}{2})$$

on the unit sphere centered at $(0,0,0)$ in $\mathbf{R}^3$. Find the vertices of the dual triangle of $\triangle ABC$.

b) By direct computation, verify Proposition 2.5.1 and Corollaries 2.5.1 and 2.5.3 for $\triangle ABC$ and its dual.

c) Sketch $\triangle ABC$ and $^*\triangle ABC$ on the unit sphere.

2.6 The Law of Cosines for Angles

We get the next result for free from the law of cosines for sides by exploiting duality.

Corollary 2.6.1 The Law of Cosines for Angles.
  Let $\triangle ABC$ be a spherical triangle with sides a opposite A, b opposite B, and c opposite C. Then

$$\cos \angle A = -\cos \angle B \cos \angle C + \sin \angle B \sin \angle C \cos \angle a. \qquad (2.18)$$

Proof. Apply the law of cosines for sides (Proposition 2.4.1) to the dual triangle $\triangle A^* B^* C^* =^* \triangle ABC$:

$$\cos \angle a^* = \cos \angle b^* \cos \angle c^* + \sin \angle b^* \sin \angle c^* \cos \angle A^*, \qquad (2.19)$$

where a^* is the side opposite A^*, b^* is opposite B^*, and c^* is opposite C^*. By Proposition 2.5.3 $\angle a^* = 180^\circ - \angle A$, $\angle b^* = 180^\circ - \angle B$, and $\angle c^* = 180^\circ - \angle C$. Plug these into Equation 2.19, then use the fact that $\cos(180^\circ - \theta) = -\cos\theta$ and $\sin(180^\circ - \theta) = \sin\theta$ for every angle θ to get Equation 2.18.

This completes the proof.

Given its vertex angles one can completely solve a spherical triangle by using the law of cosines—two spherical triangles with the same vertex angles are congruent. In plane geometry one can say only that two triangles with the same angles are similar.

Example 2.6.1 Find the sides $\angle a$, $\angle b$, $\angle c$ of a spherical triangle if its vertex angles are $\angle A = 60^\circ$, $\angle B = 70^\circ$, $\angle C = 80^\circ$.

Solution. By the law of cosines for angles,

$$
\begin{aligned}
\cos \angle a &= \frac{\cos \angle A + \cos \angle B \cos \angle C}{\sin \angle B \sin \angle C} \\
&\approx \frac{.50000 + (.34202)(.17365)}{(.93969)(.98481)} \\
&\approx .60447
\end{aligned}
$$

so

$$
\angle a \approx \arccos(.60447) \approx 52.809^{\circ}.
$$

Similarly

$$
\begin{aligned}
\cos \angle b &= \frac{\cos 70^{\circ} + \cos 60^{\circ} \cos 80^{\circ}}{\sin 60^{\circ} \sin 80^{\circ}} \\
&\approx \frac{.34202 + (.50000)(.17365)}{(.86603)(.98481)} \\
&\approx .50283
\end{aligned}
$$

$$
\begin{aligned}
\cos \angle c &= \frac{\cos 80^{\circ} + \cos 60^{\circ} \cos 70^{\circ}}{\sin 60^{\circ} \sin 70^{\circ}} \\
&\approx \frac{.17365 + (.50000)(.34202)}{(.86603)(.93969)} \\
&\approx .42352
\end{aligned}
$$

so

$$
\angle b \approx 59.813^{\circ} \text{ and } \angle c \approx 64.943^{\circ}.
$$

When one combines several spherical angles having the *same vertex* the angles add up in the same way as they do in the plane. For example $\angle a + \angle b + \angle C = 360^{\circ}$ in Fig. 2.13. The reason is that the spherical angle between two curves is, by definition, equal to the angle between tangents to the curves, so if several angles have the same vertex they combine as do angles between vectors in a single tangent plane.

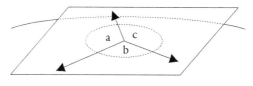

FIGURE 2.13. $a + b + c = 360^{\circ}$.

Example 2.6.2 A *dodecahedron* is a symmetrical closed surface made by joining twelve congruent regular pentagons together along their edges with three pentagons meeting at each vertex (Fig. 2.14). Find the distance from a vertex of the dodecahedron to its center if each of its edges is one inch long.

Solution. Let A and B be adjacent vertices and O the center of the dodecahedron. One could find the distance OA by solving the isosceles triangle $\triangle AOB$ if one knew the size of $\angle AOB$ since

$$(1 \text{ in.})^2 = 2(OA)^2 - 2(OA)^2 \cos \angle AOB \qquad (2.20)$$

by the law of cosines for triangles in the plane.

We shall find $\angle AOB$ by solving a spherical triangle. Project the dodecahedron out onto its circumscribing sphere, getting a spherical surface consisting of twelve congruent "spherical pentagons". Then subdivide each spherical pentagon into five congruent isosceles spherical triangles by connecting arcs from its center to its vertices (see Figure 2.14).

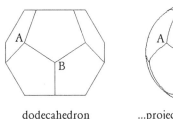
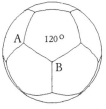
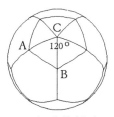

dodecahedron   ...projected onto a sphere   ...and subdivided

FIGURE 2.14.

Let $\triangle ABC$ be one of the triangles, where C is the center of a pentagon. Three spherical pentagons meet at each vertex, so each angle of the spherical pentagon measures $360^\circ/3 = 120^\circ$. Inside the pentagon two triangles meet at each vertex; it follows that

$$\angle A = \angle B = \frac{120^\circ}{2} = 60^\circ.$$

Five triangles meet at the center of the pentagon so

$$\angle C = \frac{360^\circ}{5} = 72^\circ.$$

Apply the law of cosines for angles:

$$\cos \angle \overset{\frown}{AB} = \frac{\cos 72^\circ + \cos^2 60^\circ}{\sin^2 60^\circ}$$
$$\approx 0.74536 .$$

Thus by Equation 2.20

$$OA \approx \frac{1 \text{ in.}}{2 - (2)(.74536)} \approx 1.964 \text{in.}$$

Exercise 2.6.1 ASA and AAA for Spherical Triangles.
Solve the spherical triangle $\triangle ABC$ if
a) $\angle a = 65^\circ$, $\angle b = 75^\circ$, $\angle C = 85^\circ$.
b) $\angle A = 65^\circ$, $\angle B = 75^\circ$ and $\angle C = 85^\circ$.

Exercise 2.6.2 (See Fig. 2.15). Given that its edges are one inch long, find the distance from the center to one vertex of a

a) regular tetrahedron,

b) truncated icosahedron. (A truncated icosahedron is made by joining twelve congruent regular pentagons and twenty congruent regular hexagons into a symmetrical closed surface, with one pentagon and two hexagons meeting at each vertex. It looks like a soccer ball. See Fig. 2.15.)

an icosahedron   a truncated icosahedron

FIGURE 2.15.

2.7   The Law of Sines for Spherical Triangles

Proposition 2.7.1 The Law of Sines for Spherical Triangles.

Let $\triangle ABC$ be a spherical triangle with side a opposite vertex A, side b opposite vertex B, and side c opposite vertex C. Let A^*, B^*, and C^* be the vertices of the dual triangle as in Definition 2.5.1. Then

$$\frac{\sin \angle a}{\sin \angle A} = \frac{\sin \angle b}{\sin \angle B} = \frac{\sin \angle c}{\sin \angle C}$$

$$= s \left(\frac{\det[\overrightarrow{A}, \overrightarrow{B}, \overrightarrow{C}]}{\det[\overrightarrow{A^*}, \overrightarrow{B^*}, \overrightarrow{C^*}]} \right)$$

where

$$s = \pm 1 \text{ is the sign of } \det[\overrightarrow{A}, \overrightarrow{B}, \overrightarrow{C}].$$

Proof. Let R be the radius of the sphere. By Definition 2.5.1 and Proposition 2.5.1

$$\det[\overrightarrow{A^*}, \overrightarrow{B^*}, \overrightarrow{C^*}] = \overrightarrow{A^*} \cdot (\overrightarrow{B^*} \times \overrightarrow{C^*})$$

$$= \left(\frac{\overrightarrow{B} \times \overrightarrow{C}}{R \sin \angle a} \right) \cdot \left(\overrightarrow{A} \frac{\sin a^*}{s} \right)$$

$$= \left(\frac{\sin \angle a^*}{\sin \angle a}\right) \left(\frac{\overrightarrow{A} \cdot (\overrightarrow{B} \times \overrightarrow{C})}{s}\right) \qquad (2.21)$$

$$= \left(\frac{\sin \angle a^*}{\sin \angle a}\right) \left(\frac{\det[\overrightarrow{A}, \overrightarrow{B}, \overrightarrow{C}]}{s}\right).$$

But $\sin \angle a^* = \sin \angle A$ since $\angle A + \angle a^* = 180°$. Substitute this into Equations 2.21:

$$\frac{\sin \angle a}{\sin \angle A} = \frac{\det[\overrightarrow{A}, \overrightarrow{B}, \overrightarrow{C}]}{s \det[\overrightarrow{A^*}, \overrightarrow{B^*}, \overrightarrow{C^*}]}.$$

Because $s = \pm 1$, this proves the first part of the proposition; the rest follows in a similar way.

This completes the proof.

2.8 Navigation Problems

Spherical trigonometry is commonly used to solve problems arising from astronomy and navigation.

Latitude, Longitude, and Bearings

The *latitude* of a point is the angle between the point and the equator, measured along a great circle passing through through the point and the poles. For example a point at latitude $25°$ N is $25°$ north of the equator and a point at latitude $25°$ S is $25°$ south of the equator.

The *longitude* of a point is the angle between two great circles, one connecting the point to the north and south poles and the other connecting the north and south poles to the observatory at Greenwich, England. A point at longitude $25°$ W is $25°$ west of Greenwich; a point at longitude $25°$ E is $25°$ east of Greenwich.

Bearings measure directions on the earth's surface. For example the direction bearing N $25°$ E points $25°$ to the east of due north and the direction S $25°$ W points $25°$ to the west of due south.

A *nautical mile* is the length of arc on the earth's surface that is subtended by an angle of one minute ($1' = 1/60$ degree) with vertex at the center of the earth.

A *statute mile* (5280 ft.) is the unit of distance commonly used on land. The radius of the earth is 3960 statute miles (approximately).

Thus the number of statute miles in one nautical mile is:

$$
\begin{aligned}
1 \text{ nautical mile} \ &= \ \text{distance subtended by one minute of arc} \\
&= \ \frac{1 \text{ degree}}{60} \frac{\pi \text{ radians}}{180 \text{ degrees}} (3960 \text{ statute miles}) \\
&\approx \ 1.15 \text{ statute miles}.
\end{aligned}
$$

Speed is sometimes measured in *knots*. One knot equals one nautical mile per hour.

Example 2.8.1 (Adapted from [3, page 25, example 2]). Find the distance from New Orleans to New York and the bearing (direction) from each city to the other. Use the following data:

	latitude	longitude
New Orleans	30° N	90° W
New York	41° N	74° W

Solution. Set up a spherical triangle whose three vertices are New York, New Orleans, and the north pole. Since New York is located at 41° N latitude, the angle between New York and the north pole is $90° - 41° = 49°$. The angle between New Orleans and the north pole is $90° - 30° = 60°$. The angle between the arc connecting New Orleans to the north pole and the arc connecting New York to the north pole is the difference between the longitudes of New York and New Orleans, $90° - 74° = 16°$ (Fig. 2.16).

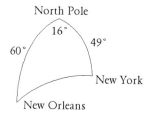

FIGURE 2.16. A geographic spherical triangle.

Let a be the arc connecting New Orleans to Boston. By the law of cosines for sides,

$$
\begin{aligned}
\cos \angle a &= \cos 49° \cos 60° + \sin 49° \sin 60° \cos 16° \\
&\approx (.65606)(.50000) + (.75471)(.86603)(.96126) \\
&\approx .95631
\end{aligned}
$$

so

$$
\begin{aligned}
\angle a &\approx 16.999° \\
&\approx .29670 \text{ radians.}
\end{aligned}
$$

Multiply this by the radius of the earth to get the distance from New York to New Orleans.

$$
\begin{aligned}
\text{distance from New York to New Orleans} &\approx (.29670)(3960 \text{ mi.}) \\
&\approx 1175 \text{ mi.}
\end{aligned}
$$

The vertex angle ∠N.O. at New Orleans measures the direction from New Orleans to New York. By the law of cosines for sides,

$$\cos \angle \text{N.O.} = \frac{\cos 49° - \cos 16.999° \cos 60°}{\sin 16.999° \sin 60°}$$

$$\approx .70266$$

so

$$\angle \text{N.O.} \approx 45.359°$$

$$\approx 45°22'.$$

Since New York lies to the east of New Orleans,

the bearing from New Orleans to New York is
N 45° 22′ E

approximately.

A similar computation shows that the vertex angle ∠N.Y at New York is N 125° 16′ approximately. According to the bearing notation we are using, all angles should measure between 0° and 90° from north or south, so this angle should be described as bearing $180° - (125°16') = 54°44'$ west from south:

the bearing from New York to New Orleans is
S 54° 44′ W

approximately.

Exercise 2.8.1 A plane flies from Los Angeles (latitude N 34°3′, longitude W 118°15′) along a great circle to Honolulu, Hawaii (latitude N 21°18′, longitude W 157°50′). In what direction is the plane headed as it leaves Los Angeles? In what direction is it headed as it approaches Honolulu? If its speed averages 500 knots (= nautical miles per hour) approximately how long will the trip take?

Exercise 2.8.2 (Adapted from [3, page 38, exercise 14]). Radio receivers at Boston, Mass., and Norfolk, N.J., detect signals from an enemy ship in a direction bearing S 83° 15′ E from Boston and N 73° 30′ E from Norfolk. Compute the latitude and longitude of the ship and its distance from Norfolk and Boston. Use the following data:

	latitude	longitude
Boston	42° 21′ N	71° 04′ W
Norfolk	36° 50′ N	76° 18′ W

(Hint. Solve a system of four spherical triangles connecting Boston, Norfolk, the ship, and the north pole. See Fig. 2.17.)

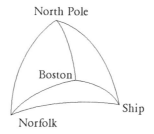

FIGURE 2.17. Four spherical triangles.

Exercise 2.8.3 The Bermuda Triangle is a region in the Atlantic Ocean where, it is said, a large number of ships and planes have disappeared for mysterious reasons. The boundaries of the Bermuda Triangle are ill-defined but for the sake of argument we shall take them to be a spherical triangle with vertices at Miami, Florida (lat. $25°$ $46'$ N, long. $80°$ $12'$ W); San Juan, Puerto Rico (lat. $18°$ $29'$ N, long. $66°$ $8'$ W); and Hamilton, Bermuda (lat. $32°$ $18'$ N, long. $64°$ $47'$ W). If a ship sinks at an unknown location in the Bermuda triangle how many square miles must be searched to find the survivors? (Hint: use Theorem 2.3.1, page 51).

2.9 Mapmaking

In Section 2.3 we proved that it is impossible to make a map of any part of a spherical earth on a flat sheet of paper without introducing some type of distortion. The mapmaker's challenge is to control the distortion so that the information he wants to depict is displayed as clearly and accurately as possible. What kind of distortion is acceptable depends on the map's intended use.

We will study four map projections:

Central Projection. A map that uses lines to represent geodesics.

Cylindrical Projections. A map that preserves areas.

Mercator Projections and Stereographic Projections.   Maps that preserve angles.

Figures 2.18–2.21 on pages 67–68 compare maps that were prepared with these projections.

Central Projections

A central projection maps a hemisphere S' onto a plane by projecting along lines extending radially from the center O of the hemisphere. Let H be a plane that does not contain O and is parallel to the great circle

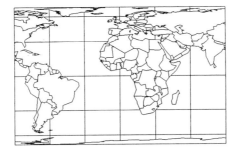

FIGURE 2.18. Cylindrical projection: latitude $S90^o$ to $N90^o$.

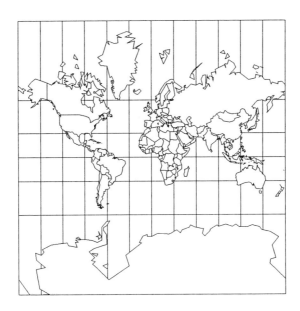

FIGURE 2.19. Mercator projection: latitude $S85^o$ to $N85^o$.

FIGURE 2.20. Central projection: latitude $N23.5^{\circ}$ to $N90^{0}$.

FIGURE 2.21. Stereographic projection: latitude $N23.5^{\circ}$ to $N90^{\circ}$.

forming the boundary of the hemisphere. The projection $f(P) \in H$ of a point $P \in S'$ is the point where the line $\overleftrightarrow{OP}$ intersects H (Fig. 2.22):

$$f(P) = \overleftrightarrow{OP} \cap H$$

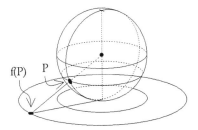

FIGURE 2.22. Central Projection.

All mappings from the sphere into the plane distort distances. With central projection the distortion is minimal near the point on S' where the tangent plane is parallel to H. Distortion becomes increasingly extreme near the edge of the hemisphere.

Proposition 2.9.1 *Let C be a curve on S'. $f(C)$ is a line segment if and only if C is an arc of a great circle (see (Fig. 2.23).*

Proof. If C is an arc of a great circle then C lies in a plane M containing O. Thus the line connecting any point on C to O lies in M. It follows that the entire projection $f(C)$ is contained in $M \cap H$. Therefore $f(C)$ is a line segment.

Conversely if $f(C)$ is a line segment then C is contained in the plane joining the line segment to the center of the sphere. Hence C is an arc of a great circle.

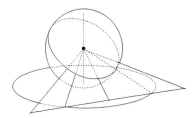

FIGURE 2.23. Projecting a great circle to a line.

As a practical application we have a method for drawing a great circle route between two points on any given map M. First locate the points on a central projection and draw a line between them. Record the the latitude and longitude coordinates of several points along this line. Now locate points with the same latitude and longitude coordinates on M and connect them with a smooth curve. The resulting curve describes a great circle route on M.

Cylindrical Projections

A cylindrical projection maps a sphere minus two diametrically opposite points into a cylinder by projecting out along lines extending radially out from a diameter of the sphere (Fig. 2.24).

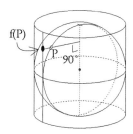

FIGURE 2.24. Cylindrical Projection.

Let S be a sphere and C a cylinder whose axis contains a diameter of S. To project a point $P \in S$ into the cylinder, extend a ray from a point on the diameter through P, perpendicular to the diameter. The projection $f(P)$ is the point where the ray intersects the cylinder (see Fig. 2.24).

Once the sphere is projected out onto the cylinder a flat map can be produced by slitting the cylinder from end to end, unrolling it, and laying it out flat. Distortions are least along the great circle where tangent planes on the sphere are parallel to tangent planes on the cylinder.

Proposition 2.9.2 *If the radii of the cylinder and sphere are equal the cylindrical projection preserves areas. In other words, if R is any region on the sphere then*

$$\text{area of } R = \text{area of } f(R).$$

Proof. It is enough to prove that

$$\text{area}(ABDC) = \text{area}(f(ABDC))$$

whenever $ABDC$ is an infinitesimal rectangle, since any area can be computed by summing up areas of infinitesimal rectangles.

Set up a system of latitude and longitude coordinates (ϕ, θ) on the surface of the sphere (see Fig. 2.25). ϕ is the latitude, measured up from the 'equator' where the cylinder is tangent to the sphere, and θ is the longitude, measured around the axis of the cylinder.[11]

Latitude and longitude meridians[12] are perpendicular to each other. Therefore the infinitesimally nearby points given in latitude and longitude coordinates by

$$A = (\phi, \theta), \qquad B = (\phi, \theta + \Delta\theta),$$
$$C = (\phi + \Delta\phi, \theta), \qquad D = (\phi + \Delta\phi, \theta + \Delta\theta),$$

are the vertices of an infinitesimal rectangle on the sphere (Fig. 2.25). Its area is

$$\text{area}(ABCD) \approx \text{length}(\overset{\frown}{AB}) \times \text{length}(\overset{\frown}{AC}), \tag{2.22}$$

and the area of its projection is

$$\text{area}(f(ABCD)) \approx \text{length}(f(\overset{\frown}{AB})) \times \text{length}(f(\overset{\frown}{AC})) \tag{2.23}$$

with the approximations becoming exact in the limit as the lengths of the sides approach zero.

Let

$$r = (\text{radius of the sphere}).$$

The arc $\overset{\frown}{AB}$ subtends an angle $\Delta\theta$ at a distance of $r\cos\phi$, measured from the axis of the cylinder. Thus

$$\text{length}(\overset{\frown}{AB}) = r\cos\phi\Delta\theta.$$

$f(\overset{\frown}{AB})$ subtends an angle of $d\theta$ at a distance r from the axis. Thus

$$\text{length}(f(\overset{\frown}{AB})) = r\,d\theta.$$

It follows that

$$\text{length}(f(\overset{\frown}{AB})) = \frac{\text{length}(\overset{\frown}{AB})}{\cos\phi}. \tag{2.24}$$

On the other hand ϕ also is the angle between the tangent to $\overset{\frown}{AC}$ at A and the cylinder, so

$$\frac{\text{length}(f(\overset{\frown}{AC}))}{\text{length}(\overset{\frown}{AC})} = \cos\phi. \tag{2.25}$$

---

[11]θ and $90^\circ - \phi$ are the spherical coordinates angles that were discussed on page 47).

[12]ϕ is constant on a latitude meridian; θ is constant on a longitude meridian.

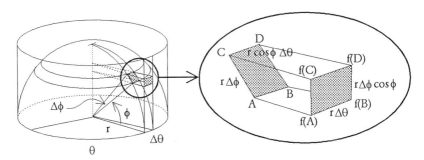

FIGURE 2.25. Equal areas.

Combining Equations 2.22–2.25, we have

$$\text{area}(f(ABDC)) \approx \text{length}(f(\overset{\frown}{AB})) \times \text{length}(f(\overset{\frown}{AC}))$$
$$\approx \text{length}(\overset{\frown}{AB}) \times \text{length}(\overset{\frown}{AC})$$
$$\approx \text{area}(ABDC)$$

with equality in the limit as the lengths approach zero.

This completes the proof.

One consequence of Proposition 2.9.2 is that anyone armed with a planimeter and a cylindrical projection can accurately measure the area of any region on the globe. Practical cylindrical projections are scaled down to human size by using a constant scale factor which must, of course, be taken into account when computing areas.

Conformal Maps

Definition 2.9.1 Conformal Mapping. A *conformal mapping* is a function f that preserves angles. f is a conformal mapping if, whenever C and C' are curves that meet in the domain of f, then the angle subtended by C and C' is congruent to the angle subtended by $f(C)$ and $f(C')$.

Example 2.9.1 An example of a conformal mapping in $\mathbf{R}^n$ is "dilation by a factor of a":

$$f : \mathbf{R}^n \rightarrow \mathbf{R}^n$$
$$f(x_1, \ldots, x_n) = (ax_1, \ldots, ax_n),$$

where $a \neq 0$ is some constant. Such a map simply scales everything up a factor of a, mapping each triangle $\triangle ABC$ to a similar triangle $f(\triangle ABC)$.

If $\triangle ABC$ is an infinitesimal triangle in the domain of a conformal map f, then $\triangle ABC$ is similar to $f(\triangle ABC)$. Thus *every conformal map acts like a dilation at the infinitesimal level.* Conversely, it can be shown that this condition is enough to ensure that f is conformal. We shall state it in the following form:

> If f acts like a dilation on infinitesimally small rectangles then
> f is a conformal map.

Conformal mappings are quite special but they are not as rare as one might suppose. For instance every complex differentiable function $f : \mathbf{C} \to \mathbf{C}$ is conformal.[13] (To see why you will have to take a course in complex analysis!)

Mercator Projections. The most common conformal map is the Mercator projection, invented by Gerhard Kremer (1512-1594), (also known as Mercator). His idea was to alter the distortion in a cylindrical projection by stretching the cylinder along its axis until the map becomes conformal. Equations 2.24 and 2.25 say that

$$\text{length}(f(\overset{\frown}{AB})) = \frac{\text{length}(\overset{\frown}{AB})}{\cos \phi}$$

while

$$\frac{\text{length}(f(\overset{\frown}{AC}))}{\text{length}(\overset{\frown}{AC})} = \cos \phi$$

so in order to make cylindrical projection conformal one needs to stretch the length of $f(\overset{\frown}{AC})$ by a factor of $1/\cos^2 \phi = \sec^2 \phi$.

Set up coordinates w and z on the surface of the cylinder, with w measuring distance around the cylinder and z measuring distance up the cylinder. When the cylinder is unrolled and laid out flat, w and z become a system of rectangular coordinates on the resulting planar map. Using elementary trigonometry one finds that if (ϕ, θ) is a point on the sphere given in latitude-longitude coordinates, then its image under cylindrical coordinate projection is the point (w, z) given by the following formulas:

$$\begin{aligned} w &= r\theta \\ z &= r\sin\phi. \end{aligned} \qquad (2.26)$$

Differentiating the second equation 2.26, one has

$$\frac{dz}{d\phi} = r\cos\phi.$$

Hence to stretch the height of an infinitesimal rectangle by a factor of $\sec^2 \phi$ one must replace z by a function v whose derivative is $\sec^2 \phi$ times as large

---

[13] $\mathbf{C}$ is the complex plane.

as the derivative of z:

$$\frac{dv}{d\phi} = r\sec^2\phi\frac{dz}{d\phi} = r\sec\phi.$$

It follows that

$$
\begin{aligned}
v &= \int \sec\phi\,d\phi \\
&= r\log|\sec\phi + \tan\phi| + C
\end{aligned}
$$

for some constant C (we shall take C to be zero). The result

$$
\begin{aligned}
w(\phi,\theta) &= a\theta, \\
v(\phi,\theta) &= a\log|\sec\phi + \tan\phi|,
\end{aligned}
$$

where a is an arbitrary positive constant, gives a formula for the Mercator projection. (Changing a merely changes the scale on the map without affecting any angles).

A line on a Mercator projection map represents a curve with a constant heading realtive to latitude and longitude lines; the angle between the curve and a longitude meridian is the same at every point on the curve. Such a curve on the earth is called a *rhumb line* or *loxodrome*; it is particularly easy for a vessel to follow since the pilot simply needs to keep a constant heading on the compass.

Exercise 2.9.1 Where do you end up if you sail forever at a constant heading (assuming that you don't run aground)?

Stereographic Projection

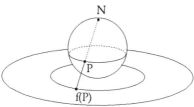

FIGURE 2.26. Stereographic Projection.

A stereographic projection maps a sphere minus one point to a plane by projecting along lines through the missing point (Fig. 2.26). Let

$$
\begin{aligned}
S &= \text{a sphere,} \\
N &= \text{a point on } S, \\
H &= \text{a plane distinct from but parallel to the} \\
&\quad\;\; \text{tangent plane at } N.
\end{aligned}
$$

If $P \in (S - \{N\})$ then its stereographic projection $f(P)$ is

$$f(P) = \overleftrightarrow{NP} \cap H.$$

Distortion is least near points on the sphere where its tangent plane is parallel to H.

To prove that stereographic projection is conformal we need two simple lemmas:

Lemma 2.9.1 *Let M_1, M_2, H_1, H_2 be planes in $\mathbf{E}^3$. If H_1 is parallel to H_2 then the lines $M_1 \cap H_1$ and $M_2 \cap H_1$ subtend the same angle as the lines $M_1 \cap H_2$ and $M_2 \cap H_2$.*

Proof. Let A be the point where the two lines meet in H_1 and B the point where the other two lines meet in H_2. Translation by $\overrightarrow{BA}$ maps one set of lines to the other (Fig. 2.27).

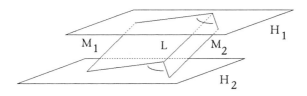

FIGURE 2.27. Equal angles.

Lemma 2.9.2 *Let C_1 and C_2 be a pair of circles in $\mathbf{E}^3$ intersecting in two points A and B. Then the circles subtend the same angle at A as they do at B.*

Proof. Reflection in the plane that forms the perpendicular bisector of $\overline{AB}$ maps the angle at A to the angle at B (Fig. 2.28).

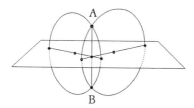

FIGURE 2.28. Equal angles.

Proposition 2.9.3 *Stereographic projection is a conformal mapping.*

Proof. (See Fig. 2.29). Let $P \neq N$ be a point on the sphere, and let $\overrightarrow{v}$, $\overrightarrow{w}$ be tangent vectors forming an angle θ at P. Let M_1 and M_2 be planes containing P and N such that

$$\begin{aligned} M_1 &= \text{ is tangent to } \overrightarrow{v}, \\ M_2 &= \text{ is tangent to } \overrightarrow{w}, \end{aligned}$$

and set

$$\begin{aligned} C_1 &= M_1 \cap S, \\ C_2 &= M_2 \cap S. \end{aligned}$$

The circles C_1 and C_2 meet at P in the same angle θ as do $\overrightarrow{v}$ and $\overrightarrow{w}$, and their projections meet in the projection of this angle. Thus we need to show that the angle formed by the circles at P is congruent to the angle formed by their projections at $f(P)$. But the angle between the circles at P is congruent to the angle between the circles at N by Lemma 2.9.2, and the angle between the circles at N is congruent to the angle between their projections by Lemma 2.9.1. Therefore the angle between the circles is congruent to the angle between their projections.

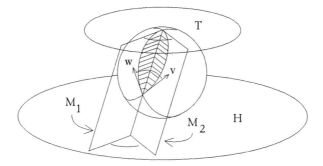

FIGURE 2.29. Stereographic projection is conformal.

Exercise 2.9.2 a) Let S be the unit sphere centered at the origin in $\mathbf{R}^3$. Show that stereographic projection of S from the point $N = (0, 0, 1)$ into the x,y plane H is given by the formula

$$f(x, y, z) = \left(\frac{x}{1-z}, \frac{y}{1-z}, 0 \right)$$

for (x, y, z) in S.

  b)Show that the inverse function is

$$f^{-1}(u, v, 0) = \left(\frac{2u}{u^2 + v^2 + 1}, \frac{2v}{u^2 + v^2 + 1}, \frac{u^2 + v^2 - 1}{u^2 + v^2 + 1} \right)$$

for $(u, v, 0)$ in the x,y plane. (Hint: find the points where the line connecting (u,v,0) to (0,0,1) intersects the sphere).

Exercise 2.9.3 Let f, S, N, and H be as in Exercise 2.9.2, and let C be a curve in S. Show that

i) C is a circle and $N \in C$ if and only if $f(C)$ is a line in H.

ii) C is a circle and $N \notin C$ if and only if $f(C)$ is a circle in H.

(Hint: a circle in S is the intersection of S with a plane $ax+by+cz+d=0$. Substitute in the formula

$$(x, y, z) = \left(\frac{2u}{u^2 + v^2 + 1}, \frac{2v}{u^2 + v^2 + 1}, \frac{u^2 + v^2 - 1}{u^2 + v^2 + 1} \right)$$

from Exercise 2.9.2 and multiply through by $u^2 + v^2 + 1$ to get a formula for the corresponding curve in H).

2.10   Applications of Stereographic Projection

Stereographic Projection in the Plane

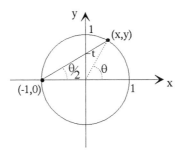

FIGURE 2.30. Stereographic Projection in the Plane.

Figure 2.30 illustrates a stereographic projection of the circle $x^2 + y^2 = 1$ in $\mathbf{R}^2$ from the point $(-1, 0)$ to the y axis. If (x, y) is a point on the circle and $(0, t)$ is its projection onto the y axis then

$$t = \frac{y}{x+1} \qquad (2.27)$$

since t is the slope of the line connecting (x, y) to $(-1, 0)$.

One can also solve for (x, y) in terms of t. Equation 2.27 says that

$$y = t(x + 1).$$

Plug this into the equation of the circle to get

$$x^2 + t^2(x + 1)^2 = 1.$$

Now subtract 1 from both sides,

$$(x^2 - 1) + t^2(x + 1)^2 = 0,$$

then factor out $(x + 1)$

$$[x + 1][(x - 1) + t^2(x + 1)] = 0.$$

Assuming $(x, y) \neq (-1, 0)$ we have

$$(x - 1) + t^2(x + 1) = 0.$$

Solve for x,

$$x = \frac{1 - t^2}{1 + t^2}$$

then plug back into the equation $y = t(x + 1)$ of the line:

$$(x, y) = \left(\frac{1 - t^2}{1 + t^2}, \frac{2t}{1 + t^2} \right). \tag{2.28}$$

Application #1. Pythagorean Triples
A *Pythagorean triple* is a sequence (X, Y, Z) of three integers such that

$$X^2 + Y^2 = Z^2. \tag{2.29}$$

For instance $(3, 4, 5)$ and $(5, 12, 13)$ are Pythagorean triples.

We shall find a formula that produces all the Pythagorean triples. If (X, Y, Z) is a Pythagorean triple then (Y, X, Z) is also, so it is enough to produce either one of them. Also if (X, Y, Z) is a Pythagorean triple and D is any integer then (DX, DY, DZ) is also a Pythagorean triple, so it is enough to find all the Pythagorean triples (X, Y, Z) where X, Y, and Z have no common factors. From now on we shall assume that X, Y, and that Z have no common factors (other than ± 1). In particular X and Y are not both zero; it follows that $Z \neq 0$, too.

Divide Equation 2.29 through by Z^2 and get

$$\frac{X^2}{Z^2} + \frac{Y^2}{Z^2} = 1. \tag{2.30}$$

Set

$$x = \frac{X}{Z} \text{ and } y = \frac{Y}{Z}. \tag{2.31}$$

x and y are nonzero rational numbers since X, Y and Z are nonzero integers. Plug Equations 2.31 into Equation 2.30 and get

$$x^2 + y^2 = 1.$$

Hence (x, y) is a point on the unit circle, with *rational* coordinates (by Equation 2.31).

Using stereographic projection, project (x, y) from $(-1, 0)$ to the y axis. By Equation 2.27 the projection of (x, y) is the point $(0, t)$ where

$$t = \frac{y}{x + 1}. \tag{2.32}$$

Hence t also is a rational number, so one can write t as a fraction reduced to lowest terms:

$$t = \frac{M}{N} \qquad (2.33)$$

where M and N are integers with no common factors (other than ± 1).

To get a formula for the Pythagorean triples we simply reverse these steps by solving for X, Y, and Z in terms of M and N. By Equation 2.28

$$(x, y) = \left(\frac{1 - t^2}{1 + t^2}, \frac{2t}{1 + t^2} \right)$$

Plug in Equations 2.33 and 2.31:

$$\left(\frac{X}{Z}, \frac{Y}{Z} \right) = \left(\frac{1 - M^2/N^2}{1 + M^2/N^2}, \frac{2M/N}{1 + M^2/N^2} \right)$$
$$= \left(\frac{N^2 - M^2}{N^2 + M^2}, \frac{2MN}{N^2 + M^2} \right) \qquad (2.34)$$

These equations are satisfied if we set

$$(X, Y, Z) = (N^2 - M^2, 2MN, M^2 + N^2). \qquad (2.35)$$

Equation 2.35 produces a Pythagorean triple for every pair of integers M and N. It remains to show that we have produced them all. The only thing that could possibly go wrong is $N^2 - M^2$, $2MN$, and $M^2 + N^2$ might all have a common factor D which cancels out in Equation 2.34, thereby leading to a solution

$$(X, Y, Z) = \left(\frac{N^2 - M^2}{D}, \frac{2MN}{D}, \frac{M^2 + N^2}{D} \right) \qquad (2.36)$$

not covered by Formula 2.35.

If P were a common *prime* factor of $N^2 - M^2$, $2MN$ and $M^2 + N^2$ then P must divide evenly into $(N^2 + M^2) + (N^2 - M^2) = 2N^2$ and also into $(N^2 + M^2) - (N^2 - M^2) = 2M^2$. It follows that either P is a common factor of M and N or else $P = 2$.

By hypothesis M and N have no common factors other than ± 1 so P must be 2. Since P divides evenly into $N^2 - M^2$ it follows that either M and N are both even numbers, or M and N are both odd numbers. In either case both $N + M$ and $N - M$ are even, so

$$M' = \frac{N + M}{2} \qquad \text{and} \qquad N' = \frac{N - M}{2}$$

are integers. Setting $D = 2$ in Equation 2.36 we have

$$\left(\frac{N^2 - M^2}{2}, \frac{2MN}{2}, \frac{M^2 + N^2}{2} \right) = (2Y', 2X', 2Z')$$

where
$$(X', Y', Z') = (N'^2 - M'^2, 2N'M', N'^2 + M'^2).$$

This equation has the same form as 2.35, except that the formulas for X and Y are switched. Therefore we have proved the following claim.

Claim 2.10.1 (X, Y, Z) *is a Pythagorean triple if and only if it can be written in the form*

$$(X, Y, Z) = (DX', DY', DZ') \text{ or } (DY', DX', DZ')$$

where

$$(X', Y', Z') = (N^2 - M^2, 2NM, N^2 + M^2)$$

and D, M, and N are integers.

Exercise 2.10.1 Find all the Pythagorean triples with $D = 1$ and $1 \leq M < N \leq 5$ (see Claim 2.10.1).

Exercise 2.10.2 Use the result of Exercise 2.9.2 to find formulas for all quadruples (W, X, Y, Z) of integers such that $W^2 + X^2 + Y^2 = Z^2$.

Application #2. Integrals of Rational Trigonometric Functions
Consider an integration problem of the form

$$\int R(\cos\theta, \sin\theta)d\theta \tag{2.37}$$

where $R(x, y)$ is a rational function (a quotient of two polynomials). Using Formula 2.28 write

$$x = \cos\theta = \frac{1 - t^2}{1 + t^2} \tag{2.38}$$

and

$$y = \sin\theta = \frac{2t}{1 + t^2} \tag{2.39}$$

where $(x, y) \neq (-1, 0)$ is an arbitrary point on the unit circle (see Fig. 2.30).

Differentiating Equation 2.39 one gets

$$
\begin{aligned}
\cos\theta d\theta &= \frac{2 - 2t^2}{(1 + t^2)^2} dt \\
&= \left(\frac{1 - t^2}{1 + t^2}\right)\left(\frac{2}{1 + t^2} dt\right) \\
&= \cos\theta \frac{2}{1 + t^2} dt
\end{aligned}
$$

where the last line follows from Equation 2.38. Thus

$$d\theta = \frac{2}{1 + t^2} dt. \tag{2.40}$$

Substitute Formulas 2.39, 2.38, and 2.40 into the integral on Line 2.37 to transform the integral into the form

$$\int R\left(\frac{1-t^2}{1+t^2}, \frac{2t}{1+t^2}\right) \frac{2}{1+t^2} dt. \tag{2.41}$$

The integrand is now a rational function of t which can be integrated by partial fractions.

Example 2.10.1 To compute

$$\int \sec\theta d\theta = \int \frac{1}{\cos\theta} d\theta,$$

substitute in Formulas 2.38, 2.39, and 2.40. The integral becomes

$$\begin{aligned}
\int \frac{1+t^2}{1-t^2} \frac{2}{1+t^2} dt &= \int \frac{2}{1-t^2} dt \\
&= \int \frac{1}{1+t} + \frac{1}{1-t} dt \text{ (by partial fractions)} \\
&= \log|1+t| - \log|1-t| + C. \tag{2.42}
\end{aligned}$$

To get a solution to the original problem, rewrite this as a function of θ. Combine Equations 2.27, 2.38, 2.39 to get

$$t = \frac{\sin\theta}{1+\cos\theta}$$

and substitute this into $\log|1+t| - \log|1-t| + C$, yielding the solution

$$\log\left|1 + \frac{\sin\theta}{1+\cos\theta}\right| - \log\left|1 - \frac{\sin\theta}{1+\cos\theta}\right| + C.$$

Alternatively, from Fig. 2.30 one can see that

$$t = \tan\frac{\theta}{2}$$

so Formula 2.42 can be written [14]

$$\log\left|1 + \tan\frac{\theta}{2}\right| - \log\left|1 - \tan\frac{\theta}{2}\right| + C.$$

A third alternative is to use properties of logarithms to rearrange Formula 2.42. Write

$$\log|1+t| - \log|1-t| = \log\left|\frac{1+t}{1-t}\right|$$

---

[14]For this reason the techniques in this section are often called "the $\tan\theta/2$ substitution" in calculus books.

$$= \log \left| \frac{1+t}{1-t} \frac{1+t}{1+t} \right|$$

$$= \log \left| \frac{1+t^2}{1-t^2} + \frac{2t}{1-t^2} \right|$$

$$= \log \left| \frac{1}{\cos \theta} + \frac{\sin \theta}{\cos \theta} \right|$$

$$= \log | \sec \theta + \tan \theta |$$

where the third line comes from Equations 2.38 and 2.39. This is the formula that one finds in calculus books.

Exercise 2.10.3 Use the techniques of this section to evaluate

$$\int \frac{1}{1 + \cos \theta} d\theta.$$

3
Conics

3.1 Conic Sections

If you take two intersecting lines L and A in $\mathbf{E}^3$ and revolve L around A then the rotating line will sweep out a *right circular cone* (unless the lines are perpendicular, in which case the rotating line sweeps out a plane). Every line that is obtained by rotating L around A is called a *generator* of the cone. The line A is the cone's *axis*, the point V where L and A intersect is the cone's *vertex*, and although this is not standard terminology, we will call the angle α between L and A the *"vertex half-angle"* of the cone $(0 < \alpha < 90^o$, see Fig. 3.1).

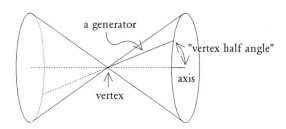

FIGURE 3.1. Right circular cone.

A *conic section* (or simply "a conic") is obtained by intersecting a right circular cone with a plane.

For the rest of this chapter we will use the following notation:

K = a right circular cone with vertex at V,

H = a plane,

$C = H \cap K$, a conic section,

α = the vertex half-angle of K,

β = the angle between H and the axis of K.

Figure 3.2 shows what α and β look like from the side, with the plane H viewed "edge-on" in various positions.

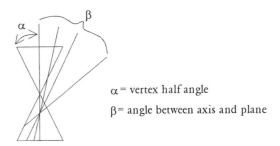

α = vertex half angle

β = angle between axis and plane

FIGURE 3.2. Side View.

The overall shape of the conic depends on two things: the relative sizes of the angles α and β, and whether or not $V \in H$. C is

    a. *smooth* if $V \notin H$,
    b. *degenerate* or, equivalently, *singular* if $V \in H$,

and it is

    i. *elliptic* if $\alpha < \beta$,
    ii. *parabolic* if $\alpha = \beta$,
    iii. *hyperbolic* if $\alpha > \beta$.

Thus there are $2 \times 3 = 6$ basic types of conics (see Fig. 3.3). A singular hyperbolic consists of two lines intersecting at V, a singular parabolic is a single line, and a singular elliptic is a point. The interesting conics are the smooth ones: the ellipse (smooth elliptic), the parabola (smooth parabolic), and the hyperbola (smooth hyperbolic).

Exercise 3.1.1 The Sundial. The common sundial consists of a horizontal plane and a vertical pointer or *gnomen*. As the sun travels across the sky it causes the shadow of the gnomen to move across the plane. The position of the shadow tells the observer the time of day.

    The tip of the moving shadow traces out a curve whose shape depends on two things: 1) the latitude of the place where the sundial is located (this determines the angle between the gnomen and the earth's axis of

Smooth Conics

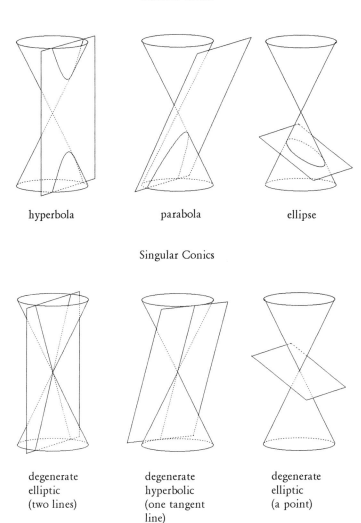

FIGURE 3.3. The six types of conics.

revolution), and 2) the season (this determines the position of the earth in its orbit around the sun, and hence the angle between the earth's axis and rays of light that come from the sun. See Fig. 3.4.)

What curve is traced out by the tip of the shadow on a sundial, and how is its shape affected by the season and the latitude? What will its shape be if the sundial is located

    a. at the north pole,
    b. at latitude 45° north,
    c. on the equator,

on

    i. the summer solstice (the day when the sun is highest in the sky),
    ii. the winter solstice (the day when the sun is lowest in the sky),
    iii. the equinox (the day when the sun passes over the equator)?

(Hint: Instead of thinking of the earth rotating around its axis, it may help to think of the earth as stationary and to regard the sun as moving around a huge circle in the sky. The size of the earth is insignificant compared to the distance to the sun so you may pretend that the earth's radius is zero. The angle between the earth's axis and the plane containing the earth's orbit around the sun is about 66.5 degrees.)

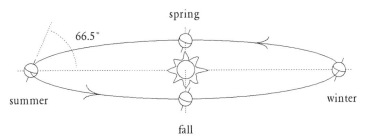

FIGURE 3.4. Four seasons.

3.2   Foci of Ellipses and Hyperbolas

The focal properties of conics are quite important in practical applications. They have been studied at least since the time of the Greek geometer Appolonius of Persa (262-190 B.C). Our discussion follows a modern line of argument due to the Belgian mathematician G. P. Dandelin in 1822.

Dandelin's constructions use spheres that are inscribed in the cone K and also tangent to the plane H.[1] If the conic is an ellipse or a hyperbola exactly two inscribed spheres are tangent to H (see Figs. 3.6 and 3.9) but if the conic is a parabola only one inscribed sphere has this property. Exercise 3.2.3 gives a construction for spheres that are inscribed in K and tangent to H; in the following we will simply assume that they exist.

Let S_1 be a sphere that is inscribed in K and tangent to H. If there are two spheres with these properties call the other one S_2. A point where S_1 or S_2 is tangent to H is a *focus* of the conic C. Set

$$F_1 = S_1 \cap H \text{ and } F_2 = S_2 \cap H.$$

Proposition 3.2.1 *If C is an ellipse then $PF_1 + PF_2$ is the same for every point $P \in C$ (Fig. 3.5).*

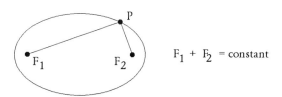

$$F_1 + F_2 = \text{constant}$$

FIGURE 3.5. Ellipse.

Proof. (See Fig. 3.6). Let P be an arbitrary point on the ellipse. $\overleftrightarrow{PF_1}$ is tangent to S_1 at F_1 and $\overleftrightarrow{PF_2}$ is tangent to S_2 at F_2 since S_1 and S_2 are tangent to H at these points. Let Q_1 and Q_2 be the circles where S_1 and S_2 intersect the cone, and set

$$R_1 = \overleftrightarrow{PV} \cap S_1 \text{ and}$$
$$R_2 = \overleftrightarrow{PV} \cap S_2.$$

Since S_1 and S_2 are tangent to K along Q_1 and Q_2, it follows that $\overleftrightarrow{PR_1}$ is tangent to S_1 at R_1 and $\overleftrightarrow{PR_2}$ is tangent to S_2 at R_2. Hence by Exercise 1.9.6,

$$PF_1 = PR_1 \text{ and } PF_2 = PR_2.$$

Therefore

$$PF_1 + PF_2 = PR_1 + PR_2.$$

But $PR_1 + PR_2 = R_1 R_2$, which is the distance between the circles Q_1 and Q_2. Since the distance between Q_1 and Q_2 does not depend on P it follows that $PF_1 + PF_2$ is the same for all $P \in C$.

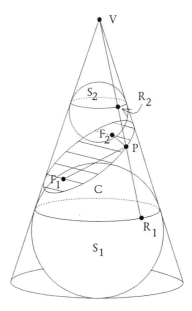

FIGURE 3.6. Dandelin's construction: ellipse.

This completes the proof.

Example 3.2.1 *An Apparatus for Drawing Ellipses.* Drive nails into the plane at F_1 and F_2. Take a piece of string whose length ℓ is greater than F_1F_2 and tie its ends to F_1 and F_2. Keeping the string stretched tight with your pencil, draw a curve around the two nails. The curve will be an ellipse with foci F_1 and F_2, satisfying the equation $PF_1 + PF_2 = \ell$ for every point P on the curve (Fig. 3.7).

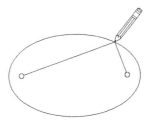

FIGURE 3.7. Apparatus for drawing ellipses.

---

[1]A sphere is inscribed in a cone if it is tangent to the cone and its center lies on the axis of the cone.

Proposition 3.2.2 *If C is a hyperbola then $|PF_1 - PF_2|$ is the same for all $P \in C$ (Fig. 3.8).*

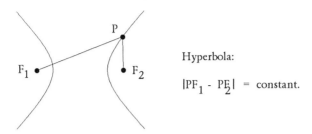

Hyperbola:

$|PF_1 - PF_2|$ = constant.

FIGURE 3.8. Hyperbola.

Exercise 3.2.1 Prove Proposition 3.2.2 by modifying the proof of Proposition 3.2.1. (Hint: compare Figs. 3.6 and 3.9).

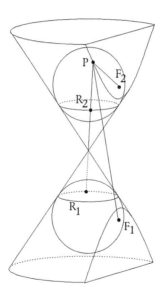

FIGURE 3.9. Dandelin's construction: hyperbola.

Example 3.2.2 *An Apparatus for Drawing Hyperbolas.* Insert nails at F_1, F_2 then attach a rod so that it pivots around the nail at F_1. Cut a piece of string with length such that

$$0 < (\text{length of rod}) - (\text{length of string}) < F_1F_2.$$

Tie one end of the string to the free end of the rod and the other end to the nail at F_2. Hold your pencil against the side of the rod and keep the string stretched tight with your pencil as you trace out a curve. The curve is part of the hyperbola that satisfies the equation $|PF_1 - PF_2| =$ (length of rod) − (length of string) for all points P on the hyperbola (Fig. 3.10).

(To get a really accurate drawing the tip of the pencil, the point where the string is attached to the rod, and the pivot F_1 should always lie in a straight line).

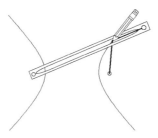

FIGURE 3.10. Apparatus for drawing hyperbolas.

Exercise 3.2.2 Show that the devices in Examples 3.2.1 and 3.2.2 perform as advertised by showing that the curves they produce satisfy the condition in Proposition 3.2.1 or Proposition 3.2.2.

Exercise 3.2.3 Show how to construct sphere(s) that are inscribed in a cone and tangent to a plane by revolving inscribed circles like those constructed in Exercise 1.9.7 on page 33 around the axis of the cone.

3.3   Eccentricity and Directrix; the Focus of a Parabola

Every noncircular smooth conic has at least one *directrix*. To construct it, let S be a circle that is inscribed in K and tangent to H. S intersects K in a circle. Every circle lies in a plane, so let E be the plane that contains $S \cap K$. The line

$$D = E \cap H$$

is a directrix of C, and the focus $F = S \cap H$ is its "associated focus" (see Fig. 3.11). If C is a circle then E is parallel to H, so the directrix does not exist.

The *eccentricity* of a conic C is the ratio

$$e = \frac{\cos \beta}{\cos \alpha}.$$

In particular,

$$\begin{aligned}
e &> 1 \text{ if } C \text{ is a hyperbola,}\\
e &= 1 \text{ if } C \text{ is a parabola,}\\
0 &< e < 1 \text{ if } C \text{ is a noncircular ellipse,}\\
e &= 0 \text{ if } C \text{ is a circle.}
\end{aligned}$$

Proposition 3.3.1 *If C is a noncircular smooth conic with eccentricity e, directrix D, and associated focus F, then*

$$PF = (e)(PD)$$

for every point $P \in C$.

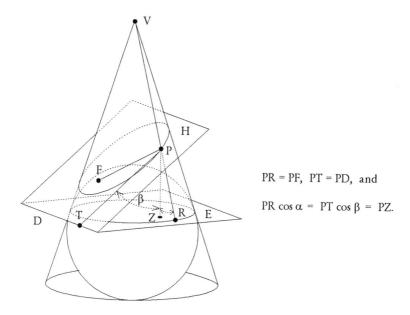

PR = PF, PT = PD, and

PR cos α = PT cos β = PZ.

FIGURE 3.11. Dandelin's construction: directrix.

Proof. (See Fig. 3.11). Let E be the plane containing $S \cap K$, and let P be an arbitrary point on C. Choose points Z, T, and R so that

$Z \in E$ and $\overline{ZP}$ is perpendicular to E,
$T \in D$ and $\overline{TP}$ is perpendicular to D, and
$R = VP \cap S$.

$\overline{ZP}$ is parallel to the axis of the cone since both $\overline{ZP}$ and the axis of the cone are perpendicular to E. Since $\overline{ZP}$ is perpendicular to E and D lies in E, it follows that $\overline{ZP}$ is perpendicular to D. Since $\overline{TP}$ is also perpendicular to D it follows that the plane $\overline{ZPT}$ is perpendicular to D. Since D lies in H,

$$\text{the plane } \overline{ZPT} \text{ is perpendicular to } H.$$

Therefore

$$\angle ZPT = \beta,$$

because $\overline{ZP}$ is parallel to the axis of the cone, and

$$\angle ZPR = \alpha$$

for the same reason. Thus

$$PZ = PR \cos \alpha = PT \cos \beta.$$

But $PR = PF$ because $\overline{PR}$ and $\overline{PF}$ are tangent to S at R and F, and $PT = PD$ because $\overline{PT}$ is perpendicular to D. Hence

$$PF \cos \alpha = PD \cos \beta.$$

Divide through by $\cos \alpha$ to complete the proof.

Corollary 3.3.1 If C is a parabola with focus F and directrix D then

$$PF = PD \text{ for all } P \in C. \tag{3.1}$$

Proof. If C is a parabola then $e = 1$. (Fig. 3.12).

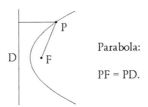

Parabola:

PF = PD.

FIGURE 3.12. Parabola.

Example 3.3.1 An Apparatus for Drawing Parabolas. Make a "T-square" by joining two rods at right angles in the shape of the letter "T". Cut a string the same length as the trunk of the "T".

Given a focus F and a directrix D, attach one end of the string to the bottom of the trunk of the "T" and the other to a point F in the plane. Trace out a curve by sliding the crosspiece of the "T" along the directrix, pulling the string tight with your pencil and keeping it in contact with the trunk of the "T". The curve is part of a parabola with focus F and directrix D (Fig. 3.13).

Exercise 3.3.1 Show that every point P on the curve traced out by the apparatus in Example 3.3.1 satisfies the equation $PF = PD$.

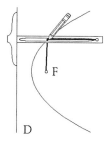

FIGURE 3.13. Apparatus for drawing parabolas.

3.4 Tangent Lines

No line can intersect a smooth conic section in more than two points. To prove this substitute the equation of the line ($x = a$ or $y = mx + b$) into the equation of the conic (see Section 3.6) and solve for x. The result is a quadratic equation

$$Ax^2 + Bx + C = 0$$

for the x coordinates of the intersection. Such equations have at most two solutions.

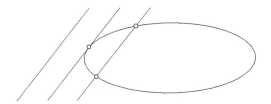

FIGURE 3.14. Lines Intersecting a Smooth Conic.

Lines that intersect the conic in exactly two points are secant lines.

A line that intersects the conic in only one point is either:

a) tangent to the conic, or

b) a line that is parallel to one of its asymptotes if the conic is a hyperbola, or

c) a line that is parallel to its axis if the conic is a parabola.

Proposition 3.4.1 *Let C be a smooth conic, P a point on C, and F a focus. If C is a parabola let D be the directrix, otherwise let $\tilde{F}$ be the other focus.*

a) If C is an ellipse then the line that bisects the angle between the vectors $-\overrightarrow{PF}$ and $\overrightarrow{P\tilde{F}}$ is tangent to C at P (Fig. 3.15).

b) If C is a hyperbola then the line that bisects the angle between $\overrightarrow{PF}$ and $\overrightarrow{P\tilde{F}}$ is tangent to C at P (Fig. 3.16).

c) If C is a parabola let $R \in D$ be the point such that $\overline{PR}$ is perpendicular to D. Then the line that bisects the angle between $\overrightarrow{PF}$ and $\overrightarrow{PR}$ is tangent to C at P (Fig. 3.17).

Proof. Let $\overleftrightarrow{AB}$ be a line through P, with A and B on opposite sides of P.

a) Assume that C is an ellipse and $\angle APF \cong \angle BP\widetilde{F}$. By Proposition 3.2.1 there exists a constant k such that

$$QF + Q\widetilde{F} = k \text{ for all } Q \in C.$$

Since $\angle APF \cong \angle BP\widetilde{F}$, Claim 1.7.1 and Exercise 1.7.1 imply that $FP\widetilde{F}$ is the shortest path from F to $\overleftrightarrow{AB}$ to $\widetilde{F}$. Hence

$$QF + Q\widetilde{F} > k \text{ if } Q \in \overleftrightarrow{AB} \text{ and } Q \neq P.$$

In other words $\overleftrightarrow{AB}$ intersects C only at P. This proves part a).

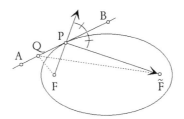

FIGURE 3.15. Tangent to an ellipse.

b) Assume that C is a hyperbola and $\angle APF \cong \angle AP\widetilde{F}$. By Proposition 3.2.2 there exists a constant k such that $|QF - Q\widetilde{F}| = k$ for all $Q \in C$.

Let E be the point on the ray $\overrightarrow{PF}$ such that $PE = P\widetilde{F}$. Then $EF = |PF - P\widetilde{F}|$, so

$$EF = k.$$

Since $\overleftrightarrow{AB}$ bisects $\angle FP\widetilde{F}$ it follows that $\overleftrightarrow{AB}$ is the perpendicular bisector of the base of the isosceles triangle $\triangle EP\widetilde{F}$. Thus if $Q \neq P$ is any other point on $\overleftrightarrow{AB}$ then $\triangle EQ\widetilde{F}$ also is isosceles, in particular

$$QE = Q\widetilde{F}.$$

Clearly $QE < QF + EF$ and $QF < QE + EF$ since Q, E and F are not collinear. Hence

$$QE - EF < QF < QE + EF.$$

Subtract QE and get $-EF < QF - QE < EF$, so

$$|QF - QE| < EF.$$

But $QE = Q\widetilde{F}$ and $EF = k$, so

$$|QF - Q\widetilde{F}| < k \qquad (3.2)$$

for all $Q \neq P$ in $\overleftrightarrow{AB}$.

This almost proves that $\overleftrightarrow{AB}$ is tangent to C; it only remains to eliminate the possibility that $\overleftrightarrow{AB}$ is parallel to an asymptotic line. The hyperbola divides the plane into three connected open regions: a middle region between the two branches of the hyperbola, and two outer regions, one containing F and the other containing $\widetilde{F}$. Define a function f on the plane by $f(Q) = QF - Q\widetilde{F}$. $f = \pm k$ on the hyperbola, and it is easy to see that $f < -k$ on the region containing F, $-k < f < k$ on the middle region, and $f > k$ on the region containing $\widetilde{F}$. Equation 3.2 says that $-k \leq f \leq k$ everywhere on $\overleftrightarrow{AB}$, so $\overleftrightarrow{AB}$ never crosses over the hyperbola into either of the outer regions. It follows that $\overleftrightarrow{AB}$ is tangent to the hyperbola.

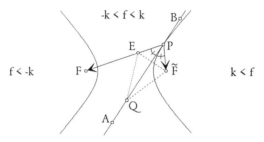

FIGURE 3.16. Tangent to a hyperbola.

c) Assume C is a parabola and $\angle FPA \cong \angle RPA$. By Corollary 3.3.1

$$Q \in C \text{ if and only if } QF = QD.$$

Since $\overline{PR}$ is perpendicular to D we have

$$PR = PF,$$

so $\triangle RPF$ is an isosceles triangle. Let $Q \in \overleftrightarrow{AB}$. $\triangle RQF$ also is an isosceles triangle, in particular

$$QF = QR.$$

If $Q \neq P$ then $\overline{QR}$ is not perpendicular to D, so $QR > QD$. Hence

$$QF - QD = QR - QD > 0.$$

It follows that $\overleftrightarrow{AB}$ intersects the parabola only at P.

The parabola bounds two open, connected regions in the plane, one containing F and the other containing D. Define a function f by $f(Q) =$

$QF - QD$. $f < 0$ on the region containing the focus and $f > 0$ on the region containing the directrix. Since $f \geq 0$ everywhere on $\overleftrightarrow{AB}$ it follows that $\overleftrightarrow{AB}$ never crosses over the parabola from one region to the other. Therefore $\overleftrightarrow{AB}$ is a tangent line.

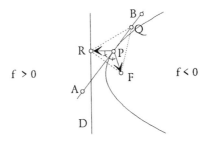

FIGURE 3.17. Tangent to a parabola.

Exercise 3.4.1 Prove that if an ellipse and a hyperbola have the same foci then they are perpendicular to each other at each point where they intersect.

3.5   Focusing Properties of Conics

Light and sound reflect off smooth curved surfaces in the same direction as they would reflect off a plane that is tangent to the surface, following the rule that *the angle of incidence equals the angle of reflection*. Because of the special relation between the foci of smooth conics and their tangents, mirrors formed by revolving an ellipse or a hyperbola around the line through its foci, or a parabola around its axis of symmetry, have unique focusing properties that are useful in applications.

Parabolic Mirrors. Figure 3.18 shows that light entering a parabolic mirror in a direction parallel to its axis of symmetry will reflect into the focus of the parabola. $\angle LPB \cong \angle RPA$ since they are "vertical angles", and $\angle RPA \cong \angle FPA$ by Proposition 3.4.1. Hence $\angle LPB = \angle FPA$, which says that the angle of incidence equals the angle of reflection.

Reflecting telescopes use parabolic mirrors because the ability of parabolic mirrors to gather a great deal of light into one spot enables astronomers to see objects that otherwise would be too dim to detect. Parabolic mirrors are also used in solar collectors, long distance microphones, and receiving antennas.

Energy radiating out from the focus of a parabolic mirror reflects into a beam that is parallel to the axis, which makes parabolic reflectors ideal for constructing headlights, spotlights, and directional transmitting antennas.

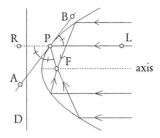

FIGURE 3.18. Parabolic reflector.

Hyperbolic Mirrors. Light aimed at one focus of a hyperbolic mirror reflects off the mirror toward the other focus. $\angle LPB \cong \angle \tilde{F}PA \cong \angle FPA$ in Figure 3.19, by Proposition 3.4.1 and the congruence of vertical angles.

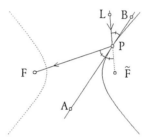

FIGURE 3.19. Hyperbolic reflector.

Some reflecting telescopes use a secondary hyperbolic mirror in addition to the main parabolic reflector to redirect the light from the main focus to a more convenient point. Both the parabola and the hyperbola in Fig. 3.20 have the same focus $\tilde{F}$. Light entering the parabolic mirror reflects toward $\tilde{F}$, then bounces off the hyperbolic mirror and travels toward the other focus of the hyperbola.

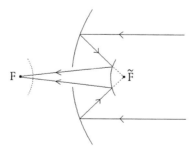

FIGURE 3.20. Compound parabolic-hyperbolic reflector.

Elliptic Mirrors. Another consequence of Proposition 3.4.1 is that energy radiating out from one focus of an elliptic reflector reflects toward the other focus. Dentists' lamps use elliptic reflectors to focus light at one spot in the patient's mouth (Fig. 3.21). The Capitol building in Washington, D.C. contains an elliptically shaped "whispering gallery", designed so that a whisper uttered at one focus can be heard at the other focus but not in other parts of the room.

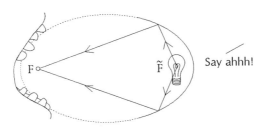

FIGURE 3.21. A dentist's elliptical lamp.

3.6   Review Exercises: Standard Equations for Smooth Conics

Exercise 3.6.1 The Equation of a Parabola.
Let C be the parabola in the x,y plane with focus $F = (c, 0)$ and directrix the line $x = c$. Using Equation 3.1 show that C satisfies the equation

$$y^2 = 4cx.$$

Exercise 3.6.2 Equations for Hyperbolas and Ellipses.
Let C be a hyperbola or ellipse in the x,y plane with foci $F_1, F_2 = (\pm c, 0)$ on the x axis, $c \geq 0$.

a) Show that there are constants $a > 0$ and $s = \pm 1$ such that

$$PF_1 + sPF_2 = \pm 2a \text{ for all } P \in C, \tag{3.3}$$

where

$$s = \begin{cases} 1 & \text{if } C \text{ is an ellipse} \\ -1 & \text{if } C \text{ is a hyperbola.} \end{cases}$$

b) Deduce that C satisfies the equation

$$\frac{x^2}{a^2} + s\frac{y^2}{b^2} = 1 \tag{3.4}$$

where

$$b^2 = s(a^2 - c^2).$$

(Hint. Rewrite Equation 3.3 in the form $PF_1 = \pm 2a - sPF_2$, substitute $P = (x, y)$, then square both sides, rearrange terms, and square again to eliminate the square roots.)

c) If C is a hyperbola show that the asymptotes have slope $\pm b/a$. (Hint: solve Equation 3.4 for y^2/x^2 then let x go to infinity.)

d) Let e be the eccentricity of the conic and D the directrix associated with the focus $(c, 0)$. Show that

$$e = \frac{c}{a}$$

and

$$D = \left\{ (x, y) | x = \frac{a}{e} \right\}.$$

(Hint. The equation says that the points $(a, 0)$ and $(-a, 0)$ lie on C. The proof of Proposition 3.3.1 shows that D is perpendicular to the line $\overleftrightarrow{F_1 F_2}$, so D satisfies an equation of the form $x = d$. Use this together with Proposition 3.3.1 to get two equations in the unknowns e and d, and solve them).

Exercise 3.6.3 Equation for a Smooth Conic in Polar Coordinates.

Let $0 < d, e$. Show that a smooth conic with focus $= (0, 0)$, associated directrix the line $x = d$, and eccentricity e is parametrized in polar coordinates by

$$r = \frac{p}{1 - e \cos \theta} \tag{3.5}$$

where

$$p = de.$$

Note that Equation 3.5 makes sense even when $e = 0$, in which case C is a circle (an ellipse with "directrix at infinity").

3.7   LORAN Navigation

Electronic Navigation. The LORAN (LOng RAnge Navigation) system enables the navigator of a ship or airplane to find its position without relying on visible landmarks. Radio stations at F_1 and F_2 simultaneously broadcast signals that are received by the ship at P. The navigator measures the interval

$$\Delta t = t_2 - t_1$$

between the time t_2 when he receives the signal sent by F_2, and the time t_1 when he receives the signal from F_1. If T_1 is the amount of time it takes the signal from F_1 to reach the ship, and T_2 is the amount of time it takes the signal from F_2 to reach the ship, then the difference between the distance from the ship to F_1 and the distance from the ship to F_2 is

$$PF_2 - PF_1 = c\Delta t$$

where c is the speed of light.

The navigator cannot measure T_1 and T_2 directly without knowing precisely when the signals were sent. But he or she can accurately measure the difference Δt between the times when the signals were received, which is enough to determine that the ship lies at some point P on the hyperbola whose equation is

$$|PF_1 - PF_2| = c\Delta t.$$

The navigator can locate the ship's position exactly if he or she receives signals from three stations F_1, F_2, F_3. Each pair of stations gives a hyperbola containing the ship, so its exact position must lie at the point where the three hyperbolas intersect. The navigator could find this position on a map by plotting the three hyperbolas and intersecting them, or by setting up coordinates and computing their intersection algebraically. (In real life it would be necessary to correct for the curvature of the earth, and to take into account the possibility that the radio signals may have been reflected and other potential sources of error.)

North

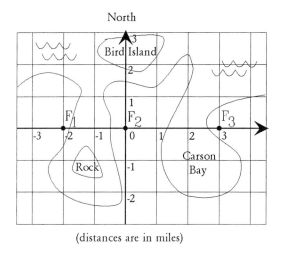

(distances are in miles)

FIGURE 3.22. Carson Bay and vicinity.

Exercise 3.7.1 An oil tanker heading for Carson Bay in dense fog receives signals that were broadcast simultaneously from three radio stations F_1, F_2, F_3, located on a line with F_2 in the middle (see Fig. 3.22). If the ship receives the signal from F_1 7.22×10^{-6} sec. later than signal from F_2, and the signal from F_3 6.30×10^{-6} sec. later than the signal from F_2, where is the ship and in which direction should the helmsman steer to avoid running aground? (The speed of light is approximately 186,000 mi./sec.)

(Hint: Set a system of coordinates with F_1, F_2, F_3 on the x axis, write equations for the hyperbolas, and solve them simultaneously).

3.8   Kepler's Laws of Planetary Motion

The elliptical shape of planetary orbits was discovered by Johannes Kepler (1571 - 1630), through careful analysis of the astronomical observations of Tycho Brahe (1546 - 1601). It is difficult to appreciate fully the magnitude of Kepler's achievement. Not only did he have to calculate planetary orbits by hand from Tycho Brahe's raw data, but he had to correct for the fact that Brahe's observations were taken from a moving platform (the earth) which was also traveling along an unknown path. And he did all this work at a time when most astronomers believed that the earth was immovably fixed at the center of the universe, with all heavenly bodies traveling in complicated paths around it.

Kepler's Laws

1. Each planet travels in an elliptical orbit with one focus at the center of mass of the system formed by the planet and the sun.
2. The vector pointing from this focus to the planet sweeps out equal areas in equal intervals of time (Fig. 3.23).
3. The cube of the period of the orbit (that is, the cube of the length of the planet's "year") is proportional to the square of the length of the orbit's major axis.

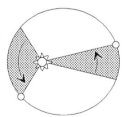

FIGURE 3.23. Equal areas in equal times.

More generally, any object in orbit around the sun travels in an orbit that has the shape of a conic section with one focus near the sun. Objects that follow closed orbits travel in ellipses; objects that are traveling fast enough to escape from the sun travel along hyperbolas or parabolas (Fig. 3.24).

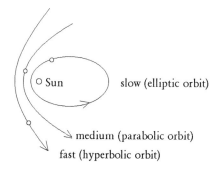

slow (elliptic orbit)

medium (parabolic orbit)

fast (hyperbolic orbit)

FIGURE 3.24. Orbits around a massive object.

Kepler's Laws provided evidence that enabled Sir Isaac Newton (1643 -1727) to formulate and confirm his famous Laws of Motion and Universal Law of Gravitation (the "inverse-square law" of gravitational force). In this section we shall reverse the historical order of events and derive Kepler's Laws from Newton's. An argument running in the opposite direction is outlined in the exercises (see exercise 3.8.1).

Newton's Laws of Motion

1. A body that is experiencing no forces moves in a straight line at a constant speed.
2. The force on a body is the product of the body's mass and its acceleration ("$f = ma$").
3. When two bodies exert forces on each other, the forces are always equal in magnitude and opposite in direction ("every action has an equal and opposite reaction").

Implicit in Newton's Laws is the assumption that the motion of the body is measured in an unaccelerated, or "inertial", coordinate system. In an inertial coordinate system, the x,y, and z axes may move through space as time passes but their acceleration is zero. One of the fundamental observations of Newtonian physics is that the physical properties of objects are the same when they are measured in an inertial coordinate system as they are in a coordinate system that is at rest. We shall elaborate on this theme in Chapter 5.

Consider an isolated system of two masses m and M, traveling freely through space, affected by no forces except each others' gravity. Let

$$\overrightarrow{r} \quad = \quad \text{the vector that points from 0 to m,} \qquad (3.6)$$

$$\overrightarrow{f} \quad = \quad \text{the force on m,}$$

$$\overrightarrow{R} \quad = \quad \text{the vector that points from 0 to M,}$$

$$\overrightarrow{F} \quad = \quad \text{the force on M,}$$

$$(3.7)$$

and let

$$r = |\overrightarrow{r}|, f = |\overrightarrow{f}|, R = |\overrightarrow{R}|, F = |\overrightarrow{F}|$$

be the magnitudes of these vectors. (Forces are represented by vectors since they have both magnitude and direction).

The center of mass of the two body system is the point

$$\text{center of mass} = m\overrightarrow{r} + M\overrightarrow{R}.$$

Newton's Second Law of Motion says that

$$\overrightarrow{f} = m\overrightarrow{r}\,'' \tag{3.8}$$
$$\overrightarrow{F} = m\overrightarrow{R}\,'',$$

where here and in the rest of this section, primes denote derivatives with respect to t = time. Newton's Third Law says that

$$\overrightarrow{F} = -\overrightarrow{f}. \tag{3.9}$$

Hence

$$m\overrightarrow{r}\,'' + M\overrightarrow{R}\,'' = 0$$

so the center of mass is not accelerating. Therefore to simplify our calculations we may change to an inertial system of coordinates whose origin is located at the center of mass. From now on we may assume that

$$m\overrightarrow{r} + M\overrightarrow{R} = 0. \tag{3.10}$$

With this assumption

$$\overrightarrow{R} = -\frac{m}{M}\overrightarrow{r} \text{ and} \tag{3.11}$$
$$\overrightarrow{F} = -\frac{m}{M}\overrightarrow{f} \tag{3.12}$$

so it is enough to calculate $\overrightarrow{r}$ and $\overrightarrow{f}$.

According to the Newtonian inverse square law of gravitation, the magnitude of the gravitational force between the two bodies is inversely proportional to the square of the distance separating the bodies and directly proportional to their masses:

$$f = \frac{GMm}{(r+R)^2}$$
$$= \frac{1}{r^2}\frac{GMm}{(1+\frac{m}{M})^2}. \tag{3.13}$$

where G is a constant (the "gravitational constant"). The second line in Equation 3.13 follows from the first by Equation 3.11. The gravitational force $\overrightarrow{f}$ on the planet m points from m towards M:

$$\text{direction of } \overrightarrow{f} = -(\text{direction of } \overrightarrow{r}). \tag{3.14}$$

To derive Kepler's second law, consider the area swept out by $\overrightarrow{r}$ as t varies over an infinitesimally small time interval Δt. The infinitesimal area ΔA swept out by $\overrightarrow{r}$ during this time interval is approximately equal to the area of the triangle whose sides are $\overrightarrow{r}$ and $\Delta \overrightarrow{r}$, or one half the area of the parallelogram whose adjacent edges are $\overrightarrow{r}$ and $\Delta \overrightarrow{r}$:

$$\Delta A \approx \frac{1}{2}|\overrightarrow{r} \times \Delta \overrightarrow{r}|,$$

where

$$\Delta \overrightarrow{r} = \overrightarrow{r}(t + \Delta t) - \overrightarrow{r}(t).$$

If Δt is very small then

$$\Delta \overrightarrow{r} \approx \overrightarrow{r}\,'\Delta t$$

so

$$\Delta A \approx \frac{1}{2}|\overrightarrow{r} \times \overrightarrow{r}\,'|\Delta t$$

with equality in the limit as $\Delta t \to 0$. Thus the area swept out by $\overrightarrow{r}$ changes at the rate of

$$\frac{dA}{dt} = \frac{1}{2}|\overrightarrow{r} \times \overrightarrow{r}\,'|. \tag{3.15}$$

Kepler's second law asserts that dA/dt is constant. To see this, differentiate $\overrightarrow{r} \times \overrightarrow{r}\,'$ in Equation 3.15,

$$\begin{aligned}
\frac{d}{dt}(\overrightarrow{r} \times \overrightarrow{r}\,') &= \overrightarrow{r}\,' \times \overrightarrow{r}\,' + \overrightarrow{r} \times \overrightarrow{r}\,'' \\
&= \overrightarrow{r} \times \overrightarrow{r}\,'' \\
&= \overrightarrow{r} \times \frac{\overrightarrow{f}}{m}
\end{aligned} \tag{3.16}$$

(the last equation comes from Newton's Second Law, Equation 3.8). Equation 3.14 says that $\overrightarrow{f}$ is parallel to $\overrightarrow{r}$. Since $\overrightarrow{r} \times \overrightarrow{r} = 0$ it follows that

$$\frac{d}{dt}(\overrightarrow{r} \times \overrightarrow{r}\,') = 0$$

Thus

$$\overrightarrow{r} \times \overrightarrow{r}\,' \text{ is a constant vector.}$$

In particular $dA/dt = (1/2)|\overrightarrow{r} \times \overrightarrow{r}\,'|$ is constant, which proves Kepler's Second Law.

Set

$$\overrightarrow{N} = \overrightarrow{r} \times \overrightarrow{r}\,'.$$ (3.17)

$\overrightarrow{r}$ points from 0 to m, and $\overrightarrow{r}$ is perpendicular to $\overrightarrow{N}$. Therefore m lies in the plane that is perpendicular to $\overrightarrow{N}$ and passes through the origin. Since $\overrightarrow{N}$ is constant this plane does not change with time, and hence it contains the entire orbit of m.

Since the orbit of m lies in a single plane we may set up coordinates in such a way that the plane containing the orbit of m is the x,y plane. Let $\overrightarrow{i}$, $\overrightarrow{j}$, and $\overrightarrow{k} = \overrightarrow{i} \times \overrightarrow{j}$ be unit vectors pointing in the x, y, and z directions, respectively. Write

$$\begin{aligned} \overrightarrow{r} &= (r\cos\theta)\overrightarrow{i} + (r\sin\theta)\overrightarrow{j} \\ &= r(\cos\theta\,\overrightarrow{i} + \sin\theta\,\overrightarrow{j}) \end{aligned}$$ (3.18)

where θ is the angle between $\overrightarrow{r}$ and the x axis. Differentiate twice with respect to t:

$$\begin{aligned} \overrightarrow{r}\,' &= r'(\cos\theta\,\overrightarrow{i} + \sin\theta\,\overrightarrow{j}) \\ &+ r\theta'(-\sin\theta\,\overrightarrow{i} + \cos\theta\,\overrightarrow{j}) \end{aligned}$$ (3.19)

and

$$\begin{aligned} \overrightarrow{r}\,'' &= (r'' - r\theta'^2)(\cos\theta\,\overrightarrow{i} + \sin\theta\,\overrightarrow{j}) \\ &+ (2r'\theta' + r\theta'')(-\sin\theta\,\overrightarrow{i} + \cos\theta\,\overrightarrow{j}), \end{aligned}$$ (3.20)

then combine Equations 3.17, 3.18, and 3.19, to get

$$(r^2\theta')\overrightarrow{k} = \overrightarrow{N}.$$

In particular, $r^2\theta'$ is constant. Setting $a = |\overrightarrow{N}|$ we have

$$\theta' = \frac{a}{r^2},$$ (3.21)

where a is a constant equal to twice the rate of change of the area swept out by $\overrightarrow{r}$.

Differentiate $a = r^2\theta'$ and get $0 = 2r'\theta' + r\theta''$. Plug this result into Equation 3.20 and get

$$\overrightarrow{r}\,'' = (r'' - r\theta'^2)(\cos\theta\,\overrightarrow{i} + \sin\theta\,\overrightarrow{j}).$$

Plug that equation into Newton's Second Law $\overrightarrow{f} = m\overrightarrow{r}\,''$, and combine with equations 3.13 and 3.14 to get a second order differential equation governing the motion of m:

$$m(r'' - r\theta'^2) = -\frac{1}{r}\frac{GMm}{(1 + \frac{m}{M})^2}.$$ (3.22)

To solve this equation we use the chain rule and Equation 3.21 to replace derivatives with respect to t with derivatives with respect to θ.

$$
\begin{aligned}
\frac{d^2 r}{dt^2} &= \frac{d}{dt}\left(\frac{dr}{d\theta}\frac{d\theta}{dt}\right) = \frac{d}{dt}\left(\frac{dr}{d\theta}\frac{a}{r^2}\right) \\
&= \frac{d}{dt}\left(\frac{dr}{d\theta}\right)\frac{a}{r^2} + \frac{dr}{d\theta}\frac{d}{dt}\left(\frac{a}{r^2}\right) \\
&= \frac{d}{d\theta}\left(\frac{dr}{d\theta}\right)\frac{d\theta}{dt}\frac{a}{r^2} + \frac{dr}{d\theta}\left(\frac{-2a}{r^3}\frac{dr}{dt}\right) \\
&= \frac{d^2 r}{d\theta^2}\left(\frac{a}{r^2}\right)^2 + \frac{dr}{d\theta}\left(\frac{-2a}{r^3}\frac{dr}{d\theta}\frac{d\theta}{dt}\right) \\
&= \frac{d^2 r}{d\theta^2}\frac{a^2}{r^4} - 2\left(\frac{dr}{d\theta}\right)^2\frac{a^2}{r^5}.
\end{aligned}
$$

Thus, by Equation 3.22,

$$
\frac{d^2 r}{d\theta^2}\frac{a^2}{r^4} - 2\left(\frac{dr}{d\theta}\right)^2\frac{a^2}{r^5} - \frac{a^2}{r^3} = -\frac{1}{r^2}\frac{GM}{(1+\frac{m}{M})^2}.
$$

Multiply through by $-r^2/a^2$ to get a second order differential equation in r and θ, governing the orbit of m.

$$
-\frac{1}{r^2}\frac{d^2 r}{d\theta^2} + \frac{2}{r^3}\left(\frac{dr}{d\theta}\right)^2 + \frac{1}{r} = \frac{GM}{a^2(1+\frac{m}{M})^2}. \tag{3.23}
$$

Equation 3.23 looks pretty nasty until you realize that it can be rewritten in the following way:

$$
\frac{d^2}{d\theta^2}\left(\frac{1}{r}\right) + \frac{1}{r} = \frac{GM}{a^2(1+\frac{m}{M})^2} \tag{3.24}
$$

a *linear* second order ordinary differential equation in $(1/r)$ and θ with constant coefficients! That is nowhere near as nasty; its general solution is

$$
\begin{aligned}
\frac{1}{r} &= \frac{GM}{a^2(1+\frac{m}{M})^2} - B\cos(\theta - \theta_0) \\
&= \frac{1 - e\cos(\theta - \theta_0)}{p} \tag{3.25}
\end{aligned}
$$

where B and θ_0 are constants, and

$$
\begin{aligned}
p &= \frac{a^2(1+\frac{m}{M})^2}{GM}, \text{ and} \tag{3.26} \\
e &= pB.
\end{aligned}
$$

Compare Equation 3.25 with Formula 3.5 on page 99, which says that the conic with focus at the origin, parameter p, eccentricity e, and directrix parallel to the y axis is given in polar coordinates by the equation

$$\frac{1}{r} = \frac{1 - e\cos\theta}{p}.$$

Equation 3.25 also describes a conic in the x,y plane with focus at the origin; the additional constant θ_0 indicates that the directrix of the conic is tilted at an angle θ_0 relative to the y axis.

This proves Kepler's first law.

It remains to prove Kepler's Third Law. If one writes t as a function of θ, then as θ runs from a to b the change in t is

$$\Delta t = \int_a^b \frac{dt}{d\theta}d\theta.$$

By Equation 3.21, $dt/d\theta = r^2/a$ so the time it takes to make one full revolution, i.e. the period, is

$$\text{period} = \int_0^{2\pi} \frac{r^2}{a}d\theta$$

if the orbit is closed. By the standard formula for areas in polar coordinates, the area inside the orbit is

$$\text{area} = \frac{1}{2}\int_0^{2\pi} r^2 d\theta,$$

so

$$\text{the period of the orbit} = \frac{2}{a}(\text{area inside the orbit}).$$

The area inside an ellipse is

$$\text{area} = \frac{\pi}{4}LL', \text{ where}$$

$$
\begin{aligned}
L &= \quad \text{(length of the major axis)}, \quad \text{and} \\
L' &= \quad \text{(length of the minor axis)}.
\end{aligned}
$$

The lengths of the major and minor axes of an ellipse with eccentricity e and parameter p are easily calculated from Formula 3.5 on page 99, using the fact that the endpoints of the major axis occur when $\theta = 0$ or π, and the endpoints of the minor axis occur where $y = r\sin\theta$ has a maximum or a minimum, that is, when $\cos\theta = e$. One finds that

$$
\begin{aligned}
L &= \frac{2p}{1 - e^2} \quad \text{and} \\
L' &= \frac{2p}{\sqrt{1 - e^2}}.
\end{aligned}
$$

Thus the area inside the orbit is

$$
\begin{aligned}
\text{area} \;&=\; \pi \frac{p^2}{(1-e^2)^{3/2}} \\
&=\; \pi p^2 \left(\frac{L}{2p}\right)^{3/2} \\
&=\; 2^{-3/2}\pi p^{1/2} L^{3/2}
\end{aligned}
$$

and the period is

$$
\begin{aligned}
\text{period} \;&=\; \frac{2}{a}(\text{area}) \\
&=\; 2^{-1/2}\pi \left(\frac{p}{a^2}\right)^{1/2} L^{3/2} \\
&=\; \pi \frac{(1+\frac{m}{M})}{\sqrt{2GM}} L^{3/2}
\end{aligned}
$$

where the last line follows from the definition of p (Equation 3.26). This proves Kepler's Third Law.

The Second Body
The orbit of the second body is

$$
\overrightarrow{R} = -\frac{m}{M}\,\overrightarrow{r}
$$

by Equation 3.11. Since they differ only by the scale factor $-m/M$, the orbits have similar shapes and equal periods.

Exercise 3.8.1 How Kepler's Laws Confirm Newton's Inverse Square Law of Gravitation. Pretend that you live at the time of Newton and you know about Kepler's Laws. Your aim is to use them to measure the gravitational force in a two-body system. Let m, $\overrightarrow{r}$, r, $\overrightarrow{f}$, f, M, $\overrightarrow{R}$, R, $\overrightarrow{F}$, F be as defined in Equations 3.6. Put the origin at the center of mass of the two-body system and treat the gravitational forces $\overrightarrow{f}$ and $\overrightarrow{F}$ as unknowns.

a) From Kepler's Second Law and the fact that the orbit of m lies in a plane, deduce that $\overrightarrow{f}$ always points directly toward or directly away from the origin, and hence that it points directly toward (or away from) M. (Hint: Use Newton"s Second Law of Motion ($\overrightarrow{f} = m\overrightarrow{r}''$) and the argument leading to Equation 3.16).

b) Deduce the inverse square law

$$
f = \frac{C}{r^2}
$$

where C is constant, from Kepler's First Law. (Hint: Let r, θ be polar coordinates in the plane containing m's orbit. Kepler's First Law says that

$$r = \frac{p}{1 - e\cos\theta}$$

for some constants p and e. The argument leading to Equation 3.22 says that

$$f = m(r'' - r\theta'^2).$$

Now use the chain rule and Equation 3.21.)

The Energy of a Two Body System
The total energy of the Keplerian two body system is

$$E = KE_m + KE_M + U$$

where

$$KE_m = \frac{1}{2}m|\vec{r}'|^2 \quad \text{and} \quad KE_M = \frac{1}{2}M|\vec{r}'|^2$$

are the kinetic energies of the bodies and

$$U = E_\infty - \frac{GMm}{r + R}$$

is the potential energy of the system. E_∞ is a constant representing the amount of energy the system would have if the bodies were infinitely far apart and were traveling at zero velocity. Using Equations 3.11, 3.19, 3.21, 3.25, and 3.26, and a little algebra, one can show that

$$E = E_\infty - \frac{1}{2p}\frac{GMm}{(1 + \frac{m}{M})}(e^2 - 1).$$

Exercise 3.8.2 a) If the orbit is closed, show that

$$E = E_\infty - \frac{1}{L}\frac{GMm}{(1 + \frac{m}{M})}$$

where L is the length of the major axis.

b) A 1000 kilogram satellite is launched into an elliptical orbit whose highest and lowest points are, respectively, 35,000 km. and 15,000 km. above the earth's surface. How much extra energy would be required to boost the satellite into a circular orbit 27,000 km. high? (Hint: If m is the mass of the satellite and M is the mass of the earth then $m/M \approx 0$. To compute GM use Newton's Second Law of Motion, the inverse square law, and the fact that the acceleration due to gravity at the earth's surface is about $9.8m/sec^2$. The mean radius of the earth is about 6371 km.)

3.9   Appendix: Reduction of a Quadratic Equation to Standard Form

Any quadratic equation

$$Ax^2 + Bxy + Cy^2 + Dx + Ey + F = 0 \qquad (3.27)$$

(where A,B,C do not all equal zero) can be reduced to one of the following standard forms by changing coordinates:

1. Standard form for a hyperbolic or a nonempty elliptic:

$$\frac{x^2}{a^2} + s\frac{y^2}{b^2} = t,$$

where

$$s = \begin{cases} 1 & \text{(the elliptic case), or} \\ -1 & \text{(the hyperbolic case),} \end{cases}$$

and

$$t = \begin{cases} 1 & \text{(the smooth case), or} \\ 0 & \text{(the degenerate case).} \end{cases}$$

2. Standard form for an empty elliptic:

$$\frac{x^2}{a^2} + \frac{y^2}{b^2} = -1.$$

3. Standard form for a parabolic:

$$y^2 = 4cx$$

where

$$c = \begin{cases} \text{a positive constant (the smooth case), or} \\ 0 \text{ (the degenerate case).} \end{cases}$$

It follows that the graph of a quadratic equation is either a conic section or the empty set.

It turns out (see Proposition 3.9.1 below) that one can determine whether the conic is hyperbolic, elliptic, parabolic, smooth, or singular without converting the equation to a standard form.

To put the Equation 3.27 in standard form, set up a new coordinate system x',y' by choosing a new origin and a new pair of perpendicular coordinate axes passing through it. If the new coordinates are chosen correctly then Equation 3.27 will be in standard form when you rewrite in the new coordinates.

In the standard system of coordinates the vector $\overrightarrow{\imath}$ points in the x direction and $\overrightarrow{\jmath}$ points in the y direction. Rotate $\overrightarrow{\imath}$ and $\overrightarrow{\jmath}$ through angle θ in the counterclockwise direction to get vectors

$$\overrightarrow{\imath}\,' = \cos\theta\,\overrightarrow{\imath} + \sin\theta\,\overrightarrow{\jmath}$$
$$\overrightarrow{\jmath}\,' = -\sin\theta\,\overrightarrow{\imath} + \cos\theta\,\overrightarrow{\jmath}.$$

Set up a new system of rectangular coordinates x',y' with origin at $(x, y) = (a, b)$ and axes pointing in the $\overrightarrow{\imath}\,'$ and $\overrightarrow{\jmath}\,'$ directions (Fig. 3.25). A point P whose coordinates are (x', y') in the new coordinate system is obtained by starting at (a, b) and moving x' units of distance in the $\overrightarrow{\imath}\,'$ direction and y' units of distance in the $\overrightarrow{\jmath}\,'$ direction:

$$P = (a, b) + x'\overrightarrow{\imath}\,' + y'\overrightarrow{\jmath}\,'.$$

Its coordinates in the original x,y coordinate system are

$$x = a + x'\cos\theta - y'\sin\theta \qquad (3.28)$$
$$y = b + x'\sin\theta + y'\cos\theta. \qquad (3.29)$$

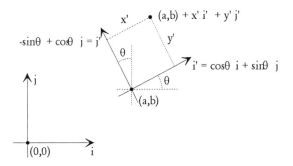

FIGURE 3.25.

After substituting Formulas 3.28 and 3.29 into Equation 3.27 we get

$$A(a + x'\cos\theta - y'\sin\theta)^2 + \cdots + F = 0.$$

Multiply this equation out and collect terms to get

$$A'x'^2 + B'x'y' + C'y'^2 + D'x' + E'y' + F' = 0 \qquad (3.30)$$

with coefficients

$$A' = A\cos^2\theta + B\cos\theta\sin\theta + C\sin^2\theta, \qquad (3.31)$$
$$B' = (C - A)\sin 2\theta + B\cos 2\theta,$$

$$\begin{aligned}
C' &= C\cos^2\theta + B\cos\theta\sin\theta + A\sin^2\theta,\\
D' &= (2Aa + Bb + D)\cos\theta + (Ba + 2Cb + E)\sin\theta,\\
E' &= (Ba + 2Cb + E)\cos\theta + (-2Aa - Bb - D)\sin\theta,\\
F' &= Aa^2 + Bab + Cb^2 + Da + Eb + F.
\end{aligned}$$

Lemma 3.9.1 *These substitutions do not change the form of the expressions*

$$B'^2 - 4A'C' = B^2 - 4AC,$$

and

$$\begin{aligned}
A'E'^2 + F'B'^2 &+ C'D'^2 - B'E'D' - 4A'C'F'\\
&= AE^2 + FB^2 + CD^2 - BED - 4ACF.
\end{aligned}$$

in Proposition 3.9.1.

Proof. (Left to the reader).

One can arrange things so that

$$B' = 0 \tag{3.32}$$

by setting

$$\theta = \frac{1}{2}\arctan\left(\frac{B}{A - C}\right).$$

If $B^2 - 4AC \neq 0$ then one can also obtain

$$D' = 0 \text{ and } E' = 0$$

by setting

$$a = \frac{2AE - BD}{B^2 - 4AC} \text{ and } b = \frac{2CD - BE}{B^2 - 4AC}.$$

In this case Equation 3.30 becomes

$$A'x'^2 + B'y'^2 + F' = 0,$$

which is easily reduced to one of the standard hyperbolic or elliptic forms.
If $B^2 - 4AC = 0$ then $B'^2 - 4A'C' = 0$ by Lemma 3.9.1, so $A' = 0$ or $C' = 0$ by Equation 3.32. Hence Equation 3.30 has the form

$$C'y'^2 + D'x' + E'y' + F' = 0$$

or

$$A'x'^2 + D'x' + E'y' + F' = 0.$$

One can eliminate the y' term in the first equation by dividing through by C', then making a further substitution of the form $y'' = (y' + E'/2C')$, $x'' = x'$ and completing the square. The second equation can be handled in a similar way by dividing through by A' then making the substitution $y'' = (x' + E'/2A')$, $x'' = y'$. In either case the result is equivalent to one of the standard elliptic equations.

Proposition 3.9.1 *If the graph of Equation 3.27 is nonempty then it is*

$$\text{hyperbolic if } B^2 - 4AC < 0,$$
$$\text{parabolic if } B^2 - 4AC = 0,$$
$$\text{elliptic if } B^2 - 4AC > 0,$$
$$\text{smooth if } AE^2 + FB^2 + CD^2 - BED - 4ACF \neq 0,$$
$$\text{singular if } AE^2 + FB^2 + CD^2 - BED - 4ACF = 0.$$

Proof. It is trivial to check that the proposition holds for a quadratic equation once it is in standard form. But by Lemma 3.9.1, changing a quadratic equation to standard form does not change the expressions $B^2 - 4AC$ or $AE^2 + FB^2 + CD^2 - BED - 4ACF$.

4
Projective Geometry

Projective geometry is an example of mathematics that was originally created with one application in mind, and yet has unexpectedly shed light on fields that are totally unrelated to the one for which it was originally developed. Created by artists during the Renaissance for analyzing perspective, projective geometry blossomed during the eighteenth and nineteenth centuries into a complete revision of the entire field of geometry. Recently it has provided the setting for the modern study of algebraic equations, and has even played a role in physics in the mathematics of quantum field theory.

4.1 Perspective Drawing

When you look at a scene your eye does not respond directly to the objects in the scene itself. It responds instead to the light rays that it receives from points in the scene. To make a correct perspective drawing the artist first *projectivizes* the scene by extending an imaginary line to his eye from each point in the scene. He then *projects* the scene into a plane by intersecting the plane with each of the imaginary lines. Taken together these intersections form an image that looks just the same as the original scene to the artist's eye (Fig. 4.1).

Projective geometry is the study of the properties of geometric figures that are not altered by projections. Two basic projections are the following (see Fig. 4.2).

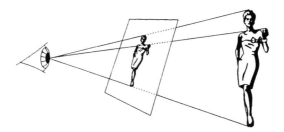

FIGURE 4.1. Projecting a scene into a plane.

1. **Central projection.** Given a point P and a plane H with $P \notin H$, define a projection function f by the formula

$$f(Q) = \overleftrightarrow{PQ} \cap H$$

for every point Q such that $\overleftrightarrow{PQ}$ is not parallel to H. f is called *projection from P into H*; P is the *center* of the projection f.

2. **Parallel projection.** Let $\overrightarrow{v}$ be a nonzero vector and H a plane that is not parallel to $\overrightarrow{v}$. For each point Q let L_Q be the line through Q that is parallel to $\overrightarrow{v}$. Define a projection function g by the formula

$$g(Q) = L_Q \cap H.$$

g is called *parallel projection into H along the direction $\overrightarrow{v}$*.

Parallel projection acts like a central projection whose center is infinitely far away.

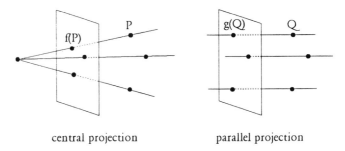

central projection   parallel projection

FIGURE 4.2.

Some examples of properties that are preserved by projections are: the property of being a point, the property of being a line, and the property of being a conic section. (Here and in the rest of this chapter we consider only objects that do not contain the artist's eye).

Properties that are not preserved include length, the size of angles, area, and the property of being a circle.

Projectivization. From now on we will imagine that the artist has only one eye, and that it is located at the origin, O. A *radial* line or plane is one that passes through O. The *projectivization* of a scene is the set of all radial lines that pass through points in the scene, together with all radial lines that are infinitesimally close to lines passing through points in the scene (nobody's eye is sharp enough to distinguish between lines that are infinitesimally close to each other).

As it views a scene your eye does not respond directly to the objects in the scene. Instead it responds to a projectivization of the scene, namely the projectivization that consists of all the light rays that travel along lines from points in the scene to your eye. This fact has important consequences:

1. **Radial lines look like points, and radial planes look like lines** because they are being viewed "edge on" by the eye at the origin.

2. **Radial dimensions are lost** because radial lines look like points.

3. **Non-radial lines acquire an extra "point at infinity".** The projectivization of a non-radial line L is the set of radial lines in the plane $\overline{OL}$ that connects L with the eye. Only one radial line in the plane $\overline{OL}$ does not connect a point on L to the eye. That one exception is the radial line that is parallel to L. We shall call this exceptional line P_∞, the "point at infinity" on L. To the eye P_∞ appears to be a point at infinity on L, because it is the limit of lines $\overleftrightarrow{OP}$ connecting the eye to points $P \in L$ as P approaches infinity (Fig. 4.3),

$$P_\infty = \lim_{P \to \infty} \overleftrightarrow{OP}.$$

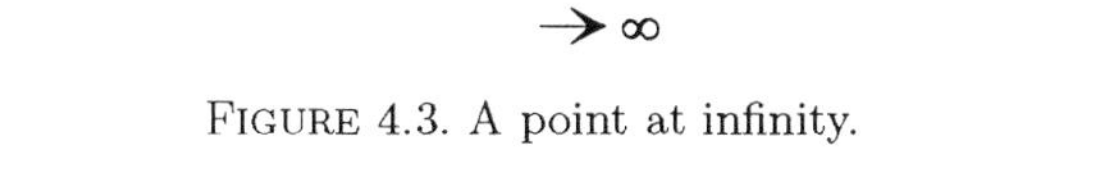

FIGURE 4.3. A point at infinity.

4. **Non-radial planes acquire an extra "line at infinity".** Let H be a non-radial plane. As $P \in H$ goes to infinity the line OP tends towards the radial plane that is parallel to H. We call this plane L_∞, the *line at infinity* of H.

$$L_\infty = \left\{ \lim_{P \to \infty} \overleftrightarrow{OP} \,\middle|\, P \in H \right\}.$$

To the eye, points in L_∞ look like they lie "at infinity" on the horizon of H (Fig. 4.4).

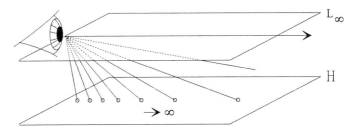

FIGURE 4.4. A line at infinity.

As a general rule any figure that extends off to infinity will acquire extra "points at infinity" when it is projectivized.

Vanishing Points.

A perspective drawing is created by intersecting the projectivization of a scene with a plane. The plane, which we shall call the *viewplane*, is the artist's canvas. The *image* of the scene is the intersection of the pprojectivization of the scene with the viewplane. A *vanishing point* is an image of a point at infinity.

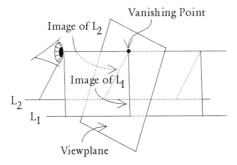

Parallel Lines have the same Vanishing Point.

FIGURE 4.5.

Parallel lines all are parallel to the same radial line, so they have the same point at infinity. Therefore the images of parallel lines all pass through the same vanishing point (Fig. 4.5). The only exception to this rule occurs when the lines are all parallel to the viewplane. In that case their point at infinity does not intersect the viewplane, so the lines have no vanishing points and their images are parallel. (Of course their common point at infinity is still visible to the eye, but it does not appear in the picture).

The image of a plane's line at infinity is the *horizon* of that plane. If the plane contains some parallel lines then their common vanishing point is on the plane's horizon.

Figure 4.6 shows four views of a rectangular box. The first is a "three point perspective"; all three vanishing points are present in the viewplane. The second view is a "two point perspective" – only two vanishing points are present. Four edges have parallel images with no vanishing point.

The third view is a "one point perspective". The viewplane is parallel to a whole side of the box.

It is impossible to have a true "zero point perspective" drawing of a rectangular box since the viewplane cannot be parallel to all its edges at once. However if you move the artist's eye off to infinity then central projection through the artist's eye becomes a parallel projection, the images of parallel lines become parallel, and the image of a box has no vanishing points. The fourth view is a parallel projection.

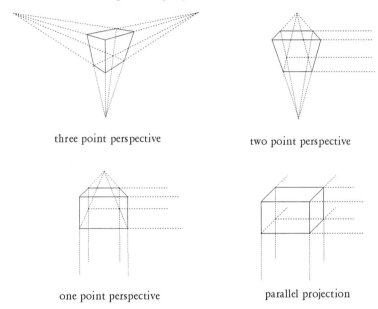

three point perspective

two point perspective

one point perspective

parallel projection

FIGURE 4.6. Four views of a box.

Parallel projections generally are preferred in technical and scientific applications even though they look less natural than true perspective drawings. Measurements in the scene and measurements in the drawing are related in a simpler way in parallel projections than in perspective drawings. This makes parallel projections easier to create and simplifies the task of calculating the exact measurements of the original object from data that are given in drawings.

Exercise 4.1.1 If you are looking at a drawing of a rectangular box in three point perspective, where should you place your eye so that the picture will look the same to you as the original box did to the artist? Let V_1, V_2, and V_3 be the three vanishing points. Show that you should put your eye at a point P such that the lines $\overleftrightarrow{PV_1}$, $\overleftrightarrow{PV_2}$, and $\overleftrightarrow{PV_3}$ are all perpendicular to each other. Let S_1 be the sphere with diameter $\overline{V_2V_3}$, S_2 the sphere with diameter $\overline{V_1V_3}$, and S_3 the sphere with diameter $\overline{V_1V_2}$. Show that $P \in S_1 \cap S_2 \cap S_3$. There are two points in this intersection, one on each side of the viewplane.

Where should you put your eye if the box is drawn in two point perspective? In one point perspective?

4.2   Projective Space

Definition 4.2.1 A *projective point* (notation: $\mathbf{P}^0$) is a radial line.

A *projective line* (notation: $\mathbf{P}^1$) is the set of radial lines in a radial plane.

A *projective plane* (notation $\mathbf{P}^2$) is the set of radial lines in a radial three dimensional space.

Generalizing the above, there are two equivalent definitions for projective space.

Definition 4.2.2 An *n dimensional projective space* $\mathbf{P}^n$ is the set of radial lines in $\mathbf{R}^{n+1}$.

Definition 4.2.3 *n dimensional projective space* $\mathbf{P}^n$ is obtained by starting with $\mathbf{R}^n$ and completing it by adding on its "points at infinity".

To see why the two definitions are equivalent, recall that

$$\mathbf{R}^{n+1} = \{(x_0, x_1, \cdots, x_n) \mid x_i \in \mathbf{R},\ i = 0, \cdots, n\}.$$

Regard $\mathbf{R}^n$ as the set of points in $\mathbf{R}^{n+1}$ with x_0 coordinate equal to one:

$$\mathbf{R}^n = \{(1, x_1, \cdots, x_n) \mid x_i \in \mathbf{R},\ i = 1, \cdots, n\},$$

a kind of "viewplane" in $\mathbf{R}^{n+1}$.

Set $x = (x_1, \cdots, x_n)$. Every point $(1, x) = (1, x_1, \cdots, x_n)$ in $\mathbf{R}^n$ lies on exactly one radial line, namely, the line $L_{(1,x)}$ consisting of all scalar multiples of the vector $(1, x)$:

$$L_{(1,x)} = \{(t, tx_1, \cdots, tx_n) \mid t \in \mathbf{R}\}.$$

Radial lines

$$L_{(0,x)} = \{(0, tx_1, \cdots, tx_n) \mid t \in \mathbf{R}\},$$

whose points have x_0 components equal to zero, are parallel to $\mathbf{R}^n$; they represent "points at infinity" on $\mathbf{R}^n$. Thus every radial line in $\mathbf{R}^{n+1}$ can be matched up either with a point $(1, x_1, \cdots, x_n)$ in $\mathbf{R}^n$ or with a point at infinity on $\mathbf{R}^n$. This produces a one to one correspondence

$$\{\text{radial lines in } \mathbf{R}^{n+1}\} \overset{1:1}{\leftrightarrow} \mathbf{R}^n \cup \{\text{points at infinity}\},$$

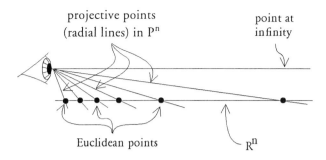

FIGURE 4.7. $\mathbf{P}^n = \mathbf{R}^n \cup \{\text{points at infinity}\}$

which shows that Definitions 4.2.2 and 4.2.3 are equivalent (see Fig. 4.7).

The fact that projective spaces contain "points at infinity" is an important difference between projective spaces and Euclidean spaces. Nevertheless the two kinds of space look the same to the eye so they usually are depicted in the same way in diagrams.

A theorem about projective space can be interpreted as a theorem about a Euclidean space of the same dimension, provided that lines meeting at a point at infinity are interpreted as parallel lines, planes meeting along a line at infinity are interpreted as parallel planes, and so on.

It should be pointed out that in projective space: *all points look exactly the same, all lines look exactly the same, and all planes look exactly the same.* In other words there is nothing special about points, lines, or planes at infinity. This means that *any* point, line, or plane in $\mathbf{P}^n$ can be regarded as a point, line, or plane at infinity, provided that this is done in a consistent way: the line at infinity in $\mathbf{P}^2$ must contain all the points at infinity, the plane at infinity in $\mathbf{P}^3$ must contain all the lines at infinity, and so on.

From now on we will drop the word "projective" wherever possible, and call projective points, projective lines and projective planes simply "points", "lines" and "planes".

Proposition 4.2.1 *(See Fig. 4.8).*

1. *Every pair of points in $\mathbf{P}^n$ lies on exactly one line.*
2. *Every pair of lines in $\mathbf{P}^2$ intersects in exactly one point.*
3. *Every pair of planes in $\mathbf{P}^3$ intersects in exactly one line.*

Proof of 1. According to Definition 4.2.2, part 1) of the proposition says that a pair of radial lines in $\mathbf{R}^{n+1}$ lies in exactly one radial plane. This is obvious; radial lines cannot be skew since they both contain the origin.

Proof of 2. Again using Definition 4.2.2, part 2) says that two radial planes in $\mathbf{R}^3$ must intersect in exactly one radial line. This is true since two radial planes cannot be parallel as both contain the origin.

Proof of 3 (sketch). A rigorous proof of part 3) requires one to think about intersecting three-dimensional radial subspaces of $\mathbf{R}^4$. Here is a more

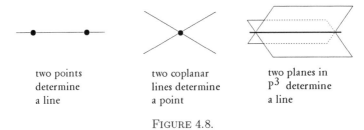

two points
determine
a line

two coplanar
lines determine
a point

two planes in
P^3 determine
a line

FIGURE 4.8.

intuitive argument. Let H and H' be planes in $\mathbf{P}^3$. Let H_∞ be a third plane in $\mathbf{P}^3$, different from the other two. Regard H_∞ as a plane at infinity, and $\mathbf{P}^3 - H_\infty$ as a copy of $\mathbf{R}^3$. The set $H - H_\infty$ of all "finite points" on H is a plane in $\mathbf{R}^3$. Likewise the set $H' - H'_\infty$ of all finite points on H' is a plane in $\mathbf{R}^3$. These two Euclidean planes either intersect in a Euclidean line or else they are parallel and have the same line at infinity. In either case it follows that H and H' intersect in a line when points at infinity are included.

Similar arguments establish the next proposition.

Proposition 4.2.2 *Any three noncollinear points in* $\mathbf{P}^3$ *lie in exactly one plane. Three planes in* $\mathbf{P}^3$ *must either intersect in exactly one point or else they contain a common line. A line and a plane in* $\mathbf{P}^3$ *intersect in exactly one point unless the line lies in the plane. If two lines in* $\mathbf{P}^3$ *do not lie in a common plane then they are skew (and do not intersect).*

4.3 Desargues' Theorem

A *triangle* $\triangle ABC$ is the union of three lines

$$\triangle ABC = \overleftrightarrow{AB} \cup \overleftrightarrow{AC} \cup \overleftrightarrow{BC}.$$

A set of lines is *coincident* if all the lines intersect at the same point. Triangles $\triangle ABC$ and $\triangle A'B'C'$ are *in perspective* if the lines $\overleftrightarrow{AA'}$, $\overleftrightarrow{BB'}$, $\overleftrightarrow{CC'}$ that join corresponding vertices on the two triangles are coincident (Fig. 4.9).

Theorem 4.3.1 (Girard Desargues, (1591-1661)).
 If the triangles $\triangle ABC$ and $\triangle A'B'C'$ are in perspective then the points

$$P = \overleftrightarrow{AB} \cap \overleftrightarrow{A'B'}, \quad Q = \overleftrightarrow{AC} \cap \overleftrightarrow{A'C'}, \quad R = \overleftrightarrow{BC} \cap \overleftrightarrow{B'C'},$$

where their corresponding sides intersect, are collinear (Fig. 4.9).

 Proof. Let

$$X = \overleftrightarrow{AA'} \cap \overleftrightarrow{BB'} \cap \overleftrightarrow{CC'}.$$

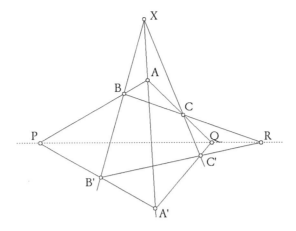

FIGURE 4.9. Desargues' Theorem.

Case 1: Suppose that $\triangle ABC$ and $\triangle A'B'C'$ do not lie in the same plane. In this case we must first check that the intersections $\overleftrightarrow{AB} \cap \overleftrightarrow{A'B'}$, $\overleftrightarrow{AC} \cap \overleftrightarrow{A'C'}$, and $\overleftrightarrow{BC} \cap \overleftrightarrow{B'C'}$ are nonempty. The fact that the triangles are in perspective implies that each of the following pairs of lines is contained in a plane:

$$\overleftrightarrow{AB}, \overleftrightarrow{A'B'} \subseteq \overline{ABX},$$
$$\overleftrightarrow{AC}, \overleftrightarrow{A'C'} \subseteq \overline{ACX},$$
$$\overleftrightarrow{BC}, \overleftrightarrow{B'C'} \subseteq \overline{BCX}.$$

Therefore each pair of lines must intersect somewhere, so the points P, Q, and R do, in fact, exist.

Clearly

$$\overleftrightarrow{AB}, \overleftrightarrow{AC}, \overleftrightarrow{BC} \subseteq \overline{ABC},$$
$$\overleftrightarrow{A'B'}, \overleftrightarrow{A'C'}, \overleftrightarrow{B'C'} \subseteq \overline{A'B'C'}.$$

Hence

$$P, Q, R \in \overline{ABC} \cap \overline{A'B'C'}$$

which is a line by part 3) of Proposition 4.2.1. This proves Desargues theorem in Case 1.

Case 2: (See Fig. 4.10). Suppose that $\triangle ABC$ and $\triangle A'B'C'$ lie in the same plane. Our approach will be to treat these triangles as projections of triangles that satisfy the conditions of Case 1.

If $A = A'$, $B = B'$, and $C = C'$ there is nothing to prove. Therefore without loss of generality we may assume that the two triangles differ on at least one edge. For the sake of argument let us assume that $\overleftrightarrow{BC} \neq \overleftrightarrow{B'C'}$.

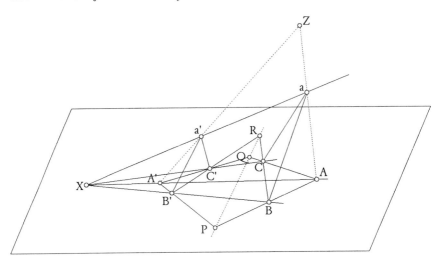

FIGURE 4.10.

Let

$$H \quad = \quad \text{a plane containing } A, B, C, A', B', C',$$
$$Z \quad \notin \quad H$$
$$a \quad \in \quad \overleftrightarrow{AZ}, \text{ with } a, A, \text{ and } Z \text{ all distinct.}$$

$\overleftrightarrow{ZA'}$ and $\overleftrightarrow{Xa}$ both lie in the plane $\overline{ZAX}$, so we can define

$$a' = \overleftrightarrow{ZA'} \cap \overleftrightarrow{Xa}.$$

By construction $X \in \overleftrightarrow{aa'} \cap \overleftrightarrow{BB'} \cap \overleftrightarrow{CC'}$ so $\triangle aBC$ and $\triangle a'B'C'$ are in perspective. Moreover they do not lie in the same plane, for H is the only plane that contains $\overleftrightarrow{BC}$ and $\overleftrightarrow{B'C'}$, but H does not contain a or a'.

By Case 1 it follows that

$$p = \overleftrightarrow{aB} \cap \overleftrightarrow{a'B'}, \quad q = \overleftrightarrow{aC} \cap \overleftrightarrow{a'C'}, \quad \text{and} \quad R$$

are collinear.

Projections map lines to lines. Since projection from Z into H maps a to A and a' to A' while fixing every point in H, it must map $p = \overleftrightarrow{aB} \cap \overleftrightarrow{a'B'}$ to $P = \overleftrightarrow{AB} \cap \overleftrightarrow{A'B'}$ and $q = \overleftrightarrow{aC} \cap \overleftrightarrow{a'C'}$ to $Q = \overleftrightarrow{AC} \cap \overleftrightarrow{A'C'}$ while leaving R fixed. Therefore since p, q and R are collinear, their images P, Q, and R must also be collinear.

This completes the proof of Desargues' theorem.

Exercise 4.3.1 Lines that Intersect at Remote Points. (Adapted from [9, Chap. 3, page 54]).

Desargues' theorem gives a way to draw a set of lines that intersect at a point outside the drawing paper. Prove that the following construction produces a line L'' that passes through a given point X and contains the intersection of two given lines L and L'.

Construction. Given two lines L and L' and a point X, draw three coincident lines M_1, M_2, and M_3 with $X \in M_1$. Let

$$
\begin{aligned}
A &= M_1 \cap L & A' &= M_1 \cap L' \\
B &= M_2 \cap L & B' &= M_2 \cap L' \\
C &= M_3 \cap L & C' &= M_3 \cap L' \\
P &= \overleftrightarrow{AB'} \cap \overleftrightarrow{A'B} & Q &= \overleftrightarrow{BC'} \cap \overleftrightarrow{B'C} \\
Y &= \overleftrightarrow{XP} \cap M_2 & Z &= \overleftrightarrow{YQ} \cap M_3.
\end{aligned}
$$

Now set

$$
L'' = \overleftrightarrow{XZ}.
$$

(Hint: $\triangle AB'C$, $\triangle A'BC'$, and $\triangle XYZ$ are in perspective. See fig. 4.11.)

The three coincident lines M_1, M_2, M_3 in exercise 4.3.1 may be parallel, that is they may meet at a point at infinity. In practice this often gives the best results.

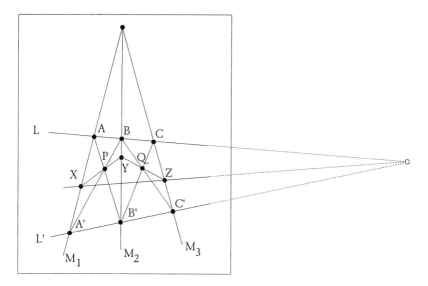

FIGURE 4.11. Lines intersecting at a remote point.

4.4 Cross Ratios

It is impossible to calculate the exact distances between objects in a scene from data obtained by measuring a perspective drawing of the scene because the drawing does not depict radial distances. However one can find the *relative* distances between three or more collinear points in the scene if they are not on the same radial line, assuming that one knows the location of the vanishing point of the line containing the points. This is accomplished by using something called the *cross ratio*.

Notation: the *ratio of two parallel vectors* in Euclidean space is

$$\frac{\overrightarrow{v}}{\overrightarrow{w}} = t \text{ if } \overrightarrow{v} = t \text{ and } \overrightarrow{w} \neq 0.$$

Definition 4.4.1 The Cross Ratio of Four Points on a Euclidean Line.

The *cross ratio* $[A, B, C, D]$ of four distinct points A, B, C, D on a Euclidean line is

$$[A, B, C, D] = \frac{\overrightarrow{AC}}{\overrightarrow{AD}} \frac{\overrightarrow{BD}}{\overrightarrow{BC}}.$$

Example 4.4.1 The cross ratio of four numbers a, b, c, d on a number line is

$$[a, b, c, d] = \frac{c - a}{d - a} \frac{d - b}{c - b}.$$

Note that if you rearrange the points their cross ratio may change. For example $[B, A, C, D] = 1/[A, B, C, D]$.

The cross ratio is significant in projective geometry because it is not changed by projections (see part 2 of the next proposition).

Proposition 4.4.1 *Let A, B, C, and D be four points on a Euclidean line, and P a point that is not on that line. Then*

1.

$$[A, B, C, D] = \frac{\sin \angle APC}{\sin \angle APD} \frac{\sin \angle BPD}{\sin \angle BPC}. \tag{4.1}$$

2. Let A', B', C', D' be the projections, respectively, of A, B, C, D from P onto another line. Then

$$[A, B, C, D] = [A', B', C', D'].$$

(See Fig. 4.12).

It is important to keep track of the signs. The angles in part 1 of the above proposition are *oriented* angles, with a direction of rotation chosen so that an angle $\angle XYZ$ indicates a rotation between $0°$ and $180°$ carrying

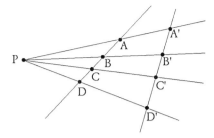

FIGURE 4.12. [A,B,C,D]=[A',B',C',D'].

$\overrightarrow{YX}$ to $\overrightarrow{YZ}$. The sines of two of these angles have the same sign if and only if the angles rotate in the same direction.

Proof. Taking into account the orientations of the angles, it is easy to check that the left and right sides of Equation 4.1 have the same signs. It remains to show that their magnitudes are equal.

The magnitude of the cross ratio is

$$
\begin{aligned}
||[A,B,C,D|| \quad &= \quad \frac{AC}{AD}\frac{BC}{BD} \\
&= \quad \frac{\text{area }\triangle\,APC}{\text{area }\triangle\,APD}\frac{\text{area }\triangle\,BPD}{\text{area }\triangle\,BPC} \\
&= \quad \left|\frac{(AP)(CP)\sin\angle APC}{(AP)(DP)\sin\angle APD}\right|\left|\frac{(BP)(DP)\sin\angle BPD}{(BP)(CP)\sin\angle BPC}\right| \\
&= \quad \left|\frac{\sin\angle APC}{\sin\angle APD}\frac{\sin\angle BPD}{\sin\angle BPC}\right|
\end{aligned}
$$

This proves part 1 of the proposition.

Part 2 is an immediate consequence of part 1 since $\angle APB = \angle A'PB'$ and so on.

This completes the proof.

If one of the four points A, B, C, D in Definition 4.4.1 is a point at infinity we can still compute the cross ratio by taking a limit. For example if $A = \infty$ then

$$[A,B,C,D] = \lim_{P\to\infty}[P,B,C,D]$$

and so on.[1] The same result is obtained simply by defining $\frac{\infty}{\infty} = 1$ and $-\frac{\infty}{\infty} = -1$.

Example 4.4.2 If b, c, d are numbers on a number line,

$$[\infty, b, c, d] \quad = \quad \lim_{p\to\infty}\frac{c-p}{d-p}\frac{d-b}{c-b}$$

---

[1] Of course P, B, C, and D must be collinear.

$$= \frac{d-b}{c-b}.$$

The conclusion of part 2 of Proposition 4.4.1 remains true if A, B, C, D or P is a point at infinity. In particular if P is a point at infinity then "projection from P" is parallel projection, and the proof of part 2 of the proposition is a simple application of similar triangles.

The next proposition shows how to compute the ratio of distances between points in a scene by using the cross ratio of their images in a perspective drawing.

Corollary 4.4.1 *Let B, C, D be three points on a Euclidean line, let B', C', D' be their images in a perspective drawing and let V' be the vanishing point of the line. Then*

$$\frac{BD}{BC} = [V', B', C', D'].$$

Proof. By Proposition 4.4.1,

$$
\begin{aligned}
[V', B', C', D'] &= [\infty, B, C, D] \\
&= \frac{BD}{BC}.
\end{aligned}
$$

Example 4.4.3 (See [9, Chap. 3, page 47.]). The following procedure will produce a perspective drawing of a Euclidean line segment that has been subdivided into n equal parts.

On the viewplane, let $\overline{AB}$ be the image of a Euclidean line segment and V be the vanishing point of the corresponding Euclidean line. Let $P \notin \overleftrightarrow{AB}$ be a point in the viewplane and let $L \neq \overrightarrow{PV}$ be a line parallel to $\overrightarrow{PV}$ in the viewplane. Set

$$A' = \overleftrightarrow{AP} \cap L \qquad \text{and} \qquad B' = \overleftrightarrow{BP}.$$

Choose points $A'_1, \ldots, A'_{n-1} \in L$ such that

$$A'A'_1 = A'_1A'_2 = \ldots = A'_{n-1}B'.$$

For each $i = 1, \ldots, n$ set

$$A_i = \overleftrightarrow{A'_iP} \cap \overline{AB}.$$

Then $A_1, A_2, \ldots, A_{n-1}$ are the images of points that subdivide the original Euclidean line segment into n segments of equal length (see Fig. 4.13).

Exercise 4.4.1 Prove that the construction in Example 4.4.3 really does produce a perspective drawing of a line segment that has been divided into pieces of equal length. (Hint: Use part 2 of Proposition 4.4.1 and Corollary 4.4.1.)

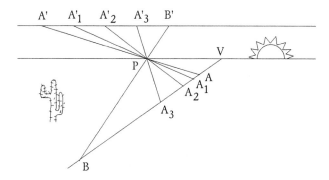

FIGURE 4.13. Equal lengths.

More Cross Ratios

Definition 4.4.1 and Proposition 4.4.1 can be extended in many ways to cover produce additional kinds of cross ratios.

Definition 4.4.2 The Cross Ratio of Four Points on a Projective Line. Let $P_1, \ldots, P_4$ be four points on a projective line, let H be a Euclidean viewplane, and let $P_1' = P \cap H, \ldots, P_4' = P_4 \cap H$ be the images of $P_1, \ldots P_4$ in H. Then

$$[P_1, P_2, P_3, P_4] = [P_1', P_2', P_3', P_4'].$$

Definition 4.4.2 would be useless if intersecting $P_1, \ldots, P_4$ with different planes H gave different cross ratios, but part 2 of Proposition 4.4.1 guarantees that this never happens.

Corollary 4.4.2 *In projective space, if $P_1, \ldots, P_4$ are four points on a line L, and $P_1', \ldots, P_4'$ are their images under a central projection mapping L to another line L', then*

$$[P_1, P_2, P_3, P_4] = [P_1', P_2', P_3', P_4'].$$

Proof. If you intersect everything with a viewplane then 4.4.2 becomes part 2 of Proposition 4.4.1.

Definition 4.4.3 The Cross Ratio of Four Coincident Lines in a Plane. Let $L_1, \ldots, L_4$ be four coincident lines in a plane. If $P_1 \in L_1$, $P_2 \in L_2$, $P_3 \in L_3$, and $P_4 \in L_4$ are any four *collinear* points, define

$$[L_1, L_2, L_3, L_4] = [P_1, P_2, P_3, P_4].$$

Definition 4.4.3 applies in both projective and Euclidean spaces. In either case, part 2 of Proposition 4.4.1 guarantees that the cross ratio of the four lines is the same regardless of the choice of the points $P_1, \ldots, P_4$.

Proposition 4.4.2 Cross ratios of lines are not changed by projections. *In* $\mathbf{P}^3$ *or* $\mathbf{E}^3$ *let* L_1, L_2, L_3, L_4 *be four coincident lines in a plane* H. *If* H' *is another plane and* f *is a central projection from* H *into* H' *then*

$$[f(L_1), f(L_2), f(L_3), f(L_4)] = [L_1, L_2, L_3, L_4].$$

Proof. (See Fig. 4.14). Let M be a line in H. For each $i = 1, \ldots, 4$ set $P_i \in M \cap L_i$. Then

$$
\begin{aligned}
[L_1, L_2, L_3, L_4] &= [P_1, P_2, P_3, P_4] \\
&= [f(P_1), f(P_2), f(P_3), f(P_4)] \\
&= [f(L_1), f(L_2), f(L_3), f(L_4)].
\end{aligned}
$$

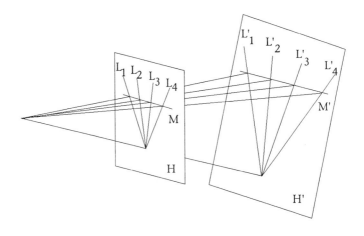

FIGURE 4.14. Cross ratios are not changed by projections.

A *conic* in the projective plane is the projectivization of a conic in the Euclidean plane. The projective conic is "smooth" if the corresponding Euclidean conic is smooth.

Proposition 4.4.3 *Let* A, B, C, D *be four points on a smooth conic* K. *Then for all* $P, Q \in K$,

$$[\overleftrightarrow{PA}, \overleftrightarrow{PB}, \overleftrightarrow{PC}, \overleftrightarrow{PD}] = [\overleftrightarrow{QA}, \overleftrightarrow{QB}, \overleftrightarrow{QC}, \overleftrightarrow{QD}].$$

Proof. (See Fig. 4.15). Clearly it is enough to prove the proposition for a Euclidean cone, for every projective cone can be made into a Euclidean cone by intersecting it with a viewplane.

Every smooth Euclidean conic is a section of a right circular cone. Let V be the vertex of the cone. If you project K from V into a plane that is perpendicular to the axis of the cone the image of K will be a circle, and the images of the lines $\overleftrightarrow{PA}, \ldots, \overleftrightarrow{QD}$ will be lines through that circle. Since cross ratios are not changed by projections the proposition is true for K if and only if it is true for the circle.

Therefore it is enough to prove the proposition in case K is a circle. Assume K is a circle. Proposition 4.4.1 and Definition 4.4.3 say that

$$[\overleftrightarrow{P'A'}, \overleftrightarrow{P'B'}, \overleftrightarrow{P'C'}, \overleftrightarrow{P'D'}] = \frac{\sin \angle A'P'C' \sin \angle B'P'D'}{\sin \angle A'P'D' \sin \angle B'P'C'}$$

and

$$[\overleftrightarrow{Q'A'}, \overleftrightarrow{Q'B'}, \overleftrightarrow{Q'C'}, \overleftrightarrow{Q'D'}] = \frac{\sin \angle A'Q'C' \sin \angle B'Q'D'}{\sin \angle A'Q'D' \sin \angle B'Q'C'}.$$

If K is a circle, Proposition 1.9.4 says that

$$\begin{array}{ll} \angle A'P'C' = \angle A'Q'C', & \angle B'P'D' = \angle B'Q'D', \\ \angle A'P'D' = \angle A'Q'D', & \angle B'P'C' = \angle B'Q'C'. \end{array}$$

It follows immediately that

$$[\overleftrightarrow{P'A'}, \overleftrightarrow{P'B'}, \overleftrightarrow{P'C'}, \overleftrightarrow{P'D'}] = [\overleftrightarrow{Q'A'}, \overleftrightarrow{Q'B'}, \overleftrightarrow{Q'C'}, \overleftrightarrow{Q'D'}].$$

This completes the proof of Proposition 4.4.3.

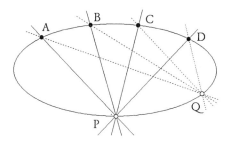

FIGURE 4.15. The cross ratios are equal.

Definition 4.4.4 The Cross Ratio of Four Points on a Smooth Conic.

The cross ratio $[A, B, C, D]$ of four points A, B, C, D on a smooth conic is defined by the formula

$$[A, B, C, D] = [\overleftrightarrow{PA}, \overleftrightarrow{PB}, \overleftrightarrow{PC}, \overleftrightarrow{PD}]$$

where P is any point on K. (By Proposition 4.4.3, $[A, B, C, D]$ is the same regardless of the point P that is used to compute it).

Exercise 4.4.2 The Complete Quadrilateral. (See [12, Chap. 4, pages 53–64]).

Let A, B, C, D be four points in a plane, no three of which lie on a line. The *complete quadrilateral ABCD* is quadrilateral $\overleftrightarrow{AB} \cup \overleftrightarrow{BC} \cup \overleftrightarrow{CD} \cup \overleftrightarrow{DA}$ together with its diagonals $\overleftrightarrow{AC}$ and $\overleftrightarrow{BD}$ (Fig. 4.16). Let

$$V_1 = \overleftrightarrow{AB} \cap \overleftrightarrow{CD}, \quad V_2 = \overleftrightarrow{AD} \cap \overleftrightarrow{BC},$$
$$V_3 = \overleftrightarrow{BD} \cap \overleftrightarrow{V_1V_2}, \quad V_4 = \overleftrightarrow{AC} \cap \overleftrightarrow{V_1V_2}.$$

Show that if one takes $\overleftrightarrow{V_1V_4}$ to be the line at infinity then A, B, C, and D are the vertices of a parallelogram in the Euclidean plane $\mathbf{P}^2 - \overleftrightarrow{V_1V_4}$. Use this fact to show that $[V_1, V_2, V_3, V_4] = -1$.

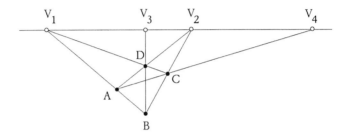

FIGURE 4.16. A complete quadrilateral.

Exercise 4.4.2 shows that every quadrilateral in $\mathbf{P}^2$ is the projectivization of a parallelogram.

The Fourth Harmonic. Given three points V_1, V_2, and V_3 on a line, draw two lines L and L' through V_1 and another line L'' through V_3. Let $B = L \cap L''$, $D = L' \cap L''$, $A = \overleftrightarrow{BV_2} \cap L'$, $C = \overleftrightarrow{DV_2} \cap L$, and $V_4 = \overleftrightarrow{AC} \cap \overleftrightarrow{V_1V_2}$.

$ABCD$ is a complete quadrilateral so $[V_1, V_2, V_3, V_4] = -1$ (see Exercise 4.4.2). V_4 is called the "fourth harmonic" of V_1, V_2, and V_3. The charming thing about this construction is that if you start with the same three points V_1, V_2, V_3 in the beginning and follow the instructions above you will always get the same point V_4 no matter what lines L, L', and L'' you use. Try it! (The reason for this is that V_4 is the only point X on $\overleftrightarrow{V_1V_2}$ that produces the cross ratio $[V_1, V_2, V_3, X] = -1$.)

One can sometimes use the diagonals of a quadrilateral instead of the technique in Example 4.4.3 to make perspective drawings of line segments of equal length. Figure 4.17 shows how to use complete quadrilaterals to draw ties on a pair of railroad tracks. The first two ties are drawn, making a quadrilateral, then the first diagonal is drawn. Next, the second diagonal is drawn with the same vanishing point as the first diagonal, and the next tie is drawn with the same vanishing point as the other ties, crossing the track at the point where the second diagonal crosses the track. This step is repeated until all the ties are drawn.

In the actual scene the rectangles formed by the tracks and the ties all are parallel and congruent. Thus their diagonals also are parallel, so corresponding sides and diagonals meet at the same points on the horizon.

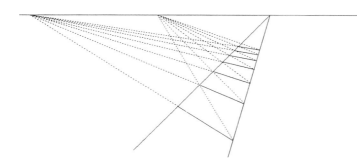

FIGURE 4.17. Railroad tracks.

Exercise 4.4.3 a) Draw an infinite checkerboard extending to the horizon in all directions.

b) Draw a scene showing a row of equally spaced houses along a street that runs toward the horizon.

4.5 Projections in Coordinates

In this section (x, y, z) are the standard coordinates in $\mathbf{R}^3$.

Example 4.5.1 Given two parallel lines

$$\begin{aligned} L_1 &= \{x = 1 \text{ and } z = -1\} \\ L_2 &= \{x = -1 \text{ and } z = -1\} \end{aligned}$$

contained in the plane
$$G = \{z = -1\},$$

project L_1 and L_2 from the origin into the viewplane

$$H = \{y = 1\}.$$

Solution. (See Fig. 4.18). Points $(x, 1, z) \in H$ and $(x', y', -1) \in G$ lie on the same radial line if and only if

$$(x', y', -1) = t(x, 1, z)$$

for some scalar t, that is, if and only if

$$x' = tx, \quad y' = t, \quad \text{and} \quad -1 = tz.$$

The third equation says that $t = -1/z$; using this the first two become

$$x' = -\frac{x}{z} \quad \text{and} \quad y' = -\frac{1}{z}. \tag{4.2}$$

$(x', y', -1)$ lies on L_1 if and only if $x' = 1$. Hence $(x, 1, z)$ lies on the projection of L_1 if and only if $-x/z = 1$. Multiplying through by z to clear the fractions, we get the equation for the projection of L_1:

$$-x = z.$$

Similarly, the equation for the projection of L_2 is $x' = -x/z = -1$, i.e.

$$x = z.$$

The horizon of G is the intersection of the viewplane with the plane $z = 0$, which is parallel to G. Thus the horizon is the line $z = 0$ in H. The projections of L_1 and L_2 meet at the common vanishing point $(x, y, z) = (0, 1, 0)$ of L_1 and L_2 in H. This vanishing point lies on the horizon of G.

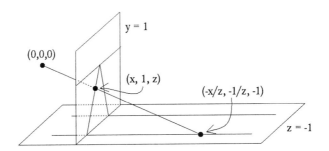

FIGURE 4.18. Images of parallel lines meeting at the horizon.

Example 4.5.2 Project the parabolas

$$Q_1 = \left\{ y = 1 + \frac{x^2}{4}, z = -1 \right\} \quad \text{and}$$

$$Q_2 = \left\{ y = \frac{x^2}{4}, z = -1 \right\}$$

from the plane $G = \{z = -1\}$ into the $y = 1$ plane.

Solution. (See Fig. 4.19). The projection of Q_1 is obtained by plugging Formulas 4.2 from the previous example into the formula for Q_1. The result is the equation $-1/z = 1 + (1/4)(-x/z)^2$. Multiply through by z^2 to clear the fractions, and get

$$z = z^2 + \frac{x^2}{4}.$$

To find out what this equation represents, complete the square then multiply by four:

$$4\left(z + \frac{1}{2}\right)^2 + x^2 = 1.$$

This is the equation of an ellipse in H with center $\left(0, 1, -\frac{1}{2}\right)$, minor axis 1 unit long, parallel to the x axis, and major axis 2 units long, parallel to the z axis. The ellipse is tangent to the horizon of G at the point $(0, 1, 0)$, which is the vanishing point of the axis of symmetry of the parabola Q_1. Thus, from the projective point of view, the parabola is simply an ellipse that is tangent to the horizon.

The image of Q_2 is computed in a similar way. Plug Formulas 4.2 into the formula for Q_2 to get $1/z = (1/4)(-x/z)^2$, then multiply through by $4z^2$ to clear the fractions:

$$4z = x^2.$$

This is the equation of a parabola that is tangent to the horizon of G at $(0, 1, 0)$. The image of Q_2 is a parabola instead of an ellipse because the radial line through the point $(0, 0, -1) \in Q_2$ is parallel to the viewplane. Thus as points on Q_2 approach $(0, 0, -1)$, their projections in the viewplane go off to infinity.

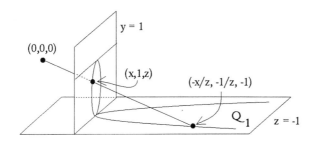

FIGURE 4.19. Projecting a parabola onto an ellipse.

Exercise 4.5.1 Project

       a) the parabola $y = -1 + z^2/4$,
       b) the circle $y^2 + z^2 = 1$,
       c) the hyperbola $y^2 - z^2 = 1$,

in the $x = 1$ plane from the origin into the $y = 1$ plane. Give a formula for the projected curve, identify what type of curve (circle, ellipse, parabola, etc.) it is, and locate the points, (if any) where it intersects the horizon of the $x = 1$ plane.

4.6   Homogeneous Coordinates and Duality

Definition 4.6.1 Homogeneous Coordinates. The *homogeneous coordinates* of a radial line in $\mathbf{R}^3$ are the Euclidean coordinates $[X, Y, Z]$ of any nonzero point on that line.

Homogeneous coordinates serve as coordinates for points on the projective plane. Note that $[X, Y, Z]$ must be nonzero. Square brackets and capital letters are used to distinguish homogeneous coordinates from Cartesian coordinates.

There is an important difference between homogeneous coordinates on $\mathbf{P}^2$ and Cartesian coordinates on $\mathbf{R}^3$: while Cartesian coordinate triples (x, y, z) are in one to one correspondence with points in $\mathbf{R}^3$, a single projective point has infinitely many sets of homogeneous coordinates.

Proposition 4.6.1 $[X, Y, Z]$ *and* $[X', Y', Z']$ *are homogeneous coordinates for the same point in* $\mathbf{P}^2$ *if and only if*

$$[X', Y', Z'] = [tX, tY, tZ]$$

for some nonzero scalar t.

Proof. (X, Y, Z) and (X', Y', Z') lie on the same radial line if and only if one of them is a scalar multiple of the other.

The Dual Projective Plane

Definition 4.6.2 The Dual Projective Plane. The *dual projective plane* $\mathbf{P}^{2*}$ is the set of all lines in $\mathbf{P}^2$. A point in $\mathbf{P}^{2*}$ is a line in $\mathbf{P}^2$.

It turns out that $\mathbf{P}^{2*}$ is a projective plane just like $\mathbf{P}^2$. To see why, recall that a line in $\mathbf{P}^2$ is a radial plane in $\mathbf{R}^3$, that is, a line is the set of all points $[X, Y, Z] \in \mathbf{P}^2$ satisfying a linear equation of the form

$$AX + BY + CZ = 0 \tag{4.3}$$

where A, B, and C are constants, at least one of them nonzero. *One can think of* $[A, B, C]$ *as a set of homogeneous coordinates for the line.* They are "homogeneous" because if t is a nonzero scalar then $[A, B, C]$ and $[tA, tB, tC]$ represent the same line: $AX + BY + CZ = 0$ if and only if $tAX + tBY + tCZ = 0$. Thus from a formal algebraic point of view, $\mathbf{P}^{2*}$ is just a projective plane whose homogeneous coordinates are represented by letters at the beginning of the alphabet instead of letters at the end.

Similarly a point in $\mathbf{P}^2$ is a line in $\mathbf{P}^{2*}$. If we regard $[A, B, C]$ as homogeneous coordinates of a point in $\mathbf{P}^{2*}$ then Equation 4.3 says that the point $[A, B, C]$ lies on the line whose coefficients are $[X, Y, Z]$. Thus $[X, Y, Z]$ becomes a line in $\mathbf{P}^{2*}$.

TABLE 4.1. Duality.

algebraic statement	interpretation	dual interpretation
$[X, Y, Z]$	point in $\mathbf{P}^2$	line in $\mathbf{P}^{2*}$
$[A, B, C]$	line in $\mathbf{P}^2$	point in $\mathbf{P}^{2*}$
$AX + BY + CZ = 0$	the point $[X, Y, Z]$ lies on the line $[A, B, C]$	the line $[X, Y, Z]$ contains the point $[A, B, C]$
$AX_1 + BY_1 + CZ_1 = 0$ and $AX_2 + BY_2 + CZ_2 = 0$	the points $[X_1, Y_1, Z_1]$ and $[X_2, Y_2, Z_2]$ lie on the line $[A, B, C]$	the lines $[X_1, Y_1, Z_1]$ and $[X_2, Y_2, Z_2]$ intersect at the point $[A, B, C]$
$A_1 X + B_1 Y + C_1 Z = 0$ and $A_2 X + B_2 Y + C_2 Z = 0$	the lines $[A_1, B_1, C_1]$ and $[A_2, B_2, C_2]$ intersect at the point $[X, Y, Z]$	the points $[A_1, B_1, C_1]$ and $[A_2, B_2, C_2]$ lie on the line $[X, Y, Z]$

The fact that there is no algebraic difference between $\mathbf{P}^2$ and $\mathbf{P}^{2*}$ besides the position of their coordinates in the alphabet, means that any statement about projective planes that can be expressed in algebraic language has two interpretations: one where "points" and "lines" are points and lines in $\mathbf{P}^2$ and a dual interpretation where "points" are points in $\mathbf{P}^{2*}$ (lines in $\mathbf{P}^2$) and "lines" are lines in $\mathbf{P}^{2*}$ (points in $\mathbf{P}^2$).

Table 4.6 summarizes this process. To dualize an existing theorem replace the words "point" by "line", "line" by "point", "intersect at" by "lie in", and "lie in" by "intersect at", wherever they occur. Since the new theorem says exactly the same thing as the old theorem when translated into algebra it is unnecessary to prove them separately. A proof of either the theorem or its dual proves them both.

Example 4.6.1 The Theorem of Pappus. (Pappus of Alexandria, circa 320 AD).

In $\mathbf{P}^2$, given points A, B, C on a line L, and A', B', C' on a line L', set

$$P = \overleftrightarrow{AB'} \cap \overleftrightarrow{A'B}, \quad Q = \overleftrightarrow{AC'} \cap \overleftrightarrow{A'C}, \quad R = \overleftrightarrow{BC'} \cap \overleftrightarrow{B'C}.$$

Then P, Q, and R are collinear (Fig. 4.20).

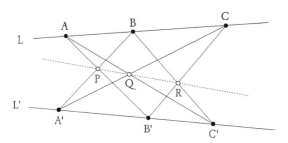

FIGURE 4.20. Theorem of Pappus.

The **Dual Theorem of Pappus.** says:

In $\mathbf{P}^2$, given lines A, B, C passing through a point L, and lines A', B', C' passing through a point L', set

$$P = \overleftrightarrow{(A \cap B')(A' \cap B)}, \quad Q = \overleftrightarrow{(A \cap C')(A' \cap C)}, \quad R = \overleftrightarrow{(B \cap C')(B' \cap C)}.$$

Then P, Q, and R are coincident (Fig. 4.21).

Exercise 4.6.1 Prove both Pappus' theorem (Example 4.6.1) and its dual by proving the dual theorem. (Hint: Regard $\mathbf{P}^2$ as the union of a Euclidean plane and its line at infinity (see page 120). Take $\overleftrightarrow{LL'}$ to be the line at infinity. Show that the lines A,B,C are parallel, and the lines A',B',C' are

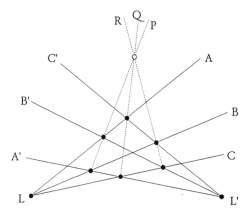

FIGURE 4.21. Dual theorem of Pappus.

also parallel, in the Euclidean plane $\mathbf{P}^2 - \overleftrightarrow{LL'}$. Write equations for these lines in x,y coordinates on the Euclidean plane, then solve for the points where they intersect).

Example 4.6.2 The **Dual of Desargues' Theorem** for triangles in $\mathbf{P}^2$ (see Proposition 4.3.1) says:

Let $\triangle ABC$ and $\triangle A'B'C'$ be triangles in $\mathbf{P}^2$ whose sides are the lines A, B, C and A', B', C', respectively. If the points $A \cap A'$, $B \cap B'$, and $C \cap C'$ all lie on the same line X, then the lines

$$P = \overleftrightarrow{(A \cap B)(A' \cap B')}, Q = \overleftrightarrow{(A \cap C)(A' \cap C')}, R = \overleftrightarrow{(B \cap C)(B' \cap C')}$$

all pass through the same point (Fig. 4.22).

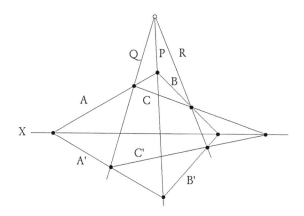

FIGURE 4.22. Dual of Desargues' theorem.

Desargues' theorem is unusual in that its dual is the same as its converse.

Exercise 4.6.2 Show that the projectivization of the line $y = mx + b$ has homogeneous coordinates $[m, -1, b]$ in $\mathbf{P}^{2^*}$.

Exercise 4.6.3 Let L_1, L_2, and L_3 be distinct, coincident lines in $\mathbf{P}^2$. Show how to construct a fourth line L_4, coincident with the other three, such that $[L_1, L_2, L_3, L_4] = -1$ (Hint: Dualize the construction of the fourth harmonic [page 132]).

4.7 Homogeneous Polynomials, Algebraic Curves

Homogeneous Polynomials Every polynomial is a sum of one or more "monomials". A monomial is a product of a constant and a finite number of variables with nonnegative integer exponents. The *degree* of a nonzero monomial is the sum of the exponents.

Example 4.7.1 z^3 *is a monomial of degree 3.*
     $2x^5 y^2 z^3$ *is a monomial of degree* $5 + 2 + 3 = 10$.

The *degree* of a polynomial is the largest of the degrees of its monomials.

Example 4.7.2 $2y^3 z^4 + 3xy + 1$ *is a polynomial of degree* $3 +$
     $4 = 7$.

A nonzero polynomial is *homogeneous* if all of its monomials have the same degree.

Example 4.7.3 $x^4 + 2yz^3$ *is a homogeneous polynomial of degree 4.*

Proposition 4.7.1 *If $f(x, y, z)$ is a homogeneous polynomial of degree d then*

$$f(tx, ty, tz) = t^d f(x, y, z)$$

for all scalars t.

Proof. If $g(x, y, z) = kx^a y^b z^c$ is a monomial of degree $a + b + c = d$ then

$$
\begin{aligned}
g(tx, ty, tz) &= k(tx)^a (ty)^b (tz)^c \\
&= kt^{a+b+c} x^a y^b z^c \\
&= t^d g(x, y, z)
\end{aligned}
$$

for every scalar t. The same holds for f since all of its terms have the same degree.
     Proposition 4.7.1 holds for polynomials in any number of variables, not just three. The proof is the same.

Corollary 4.7.1 *If f is homogeneous and $f(x, y, z) = 0$ at some nonzero point $(x, y, z) \in \mathbf{R}^3$, then $f = 0$ everywhere on the radial line through the point (x, y, z).*

Proof. The radial line consists of scalar multiples (tx, ty, tz) of (x, y, z) so the corollary follows immediately from proposition 4.7.1.

Algebraic Curves

An *algebraic curve in the Euclidean plane* is the set of points $(x, y) \in \mathbf{E}^2$ satisfying a polynomial equation

$$f(x, y) = 0$$

where f is a nonconstant polynomial. Examples include the circle $x^2 + y^2 - 1 = 0$, the line $y - 2x - 5 = 0$, the parabola $x^2 - y = 0$, and the "figure eight" $x^2(x^2 - 1) + y^2 = 0$.

An *algebraic curve in the projective plane* is the set of points $[X, Y, Z] \in \mathbf{P}^2$ satisfying a *homogeneous polynomial* equation

$$F(X, Y, Z) = 0$$

where F is a nonconstant *homogeneous* polynomial. Examples include the projectivized circle $X^2 + Y^2 - Z^2 = 0$, the projectivized line $Y - 2X - 5Z = 0$, the projectivized parabola $X^2 - YZ = 0$, and the projectivized figure eight $X^2(X^2 - Z^2) + Y^2Z^2 = 0$.

Proposition 4.7.1 says that the locus $\{(x, y, z) \in \mathbf{R}^3 | F(x, y, z) = 0\}$ of a homogeneous polynomial equation is a union of radial lines. In other words it is a kind of "generalized cone" in $\mathbf{R}^3$ with vertex at the origin. To an eye at the origin it looks like a curve (Fig. 4.23).

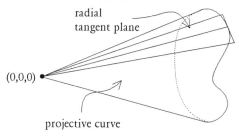

FIGURE 4.23. Generalized cone, with radial tangent plane.

Homogenization

Projectivization is the process of converting a Euclidean figure into a projective figure by replacing each of its points by a radial line. *Homogenization* is the algebraic analog of projectivization. To homogenize a polynomial $f(x_1, \ldots, x_n)$ of degree d, introduce $n + 1$ new variables $X_1, \ldots, X_n, Z$, then replace each x_i, $i = 1, \ldots, n$ by X_i/Z and multiply through by Z^d to clear fractions.

To recover the original polynomial f from the homogeneous polynomial replace each X_i by x_i and Z by 1.

Example 4.7.4 *Let $f(x,y) = 1 + x + 2y + 3x^2 + xy^3$.*

$$f\left(\frac{X}{Z}, \frac{Y}{Z}\right) = 1 + \frac{X}{Z} + 2\frac{Y}{Z} + 3\left(\frac{X}{Z}\right)^2 + \left(\frac{X}{Z}\right)\left(\frac{Y}{Z}\right)^3.$$

Multiply through by Z^4 to get the homogeneous polynomial

$$F(X, Y, Z) = Z^4 + XZ^3 + 2YZ^3 + 3X^2Z^2 + XY^3.$$

To convert F back into f set $X = x$, $Y = y$, and $Z = 1$:

$$\begin{aligned} F(x, y, 1) &= 1 + x + 2y + 3x^2 + xy^3 \\ &= f(x, y). \end{aligned}$$

Exercise 4.7.1 *Homogenize the following polynomials.*
  a) $f(t) = 1 + 2t + 3t^2 + 4t^3$.
  b) $g(x, y, z) = 1 + xyz$.

To see the relation between homogenization and projectivization, consider the Euclidean curve

$$f(x, y) = 0 \text{ and } z = 1$$

given by the polynomial equation $f(x, y) = 0$ in the $z = 1$ plane. A nonzero point (x, y, z) lies on the projectivization of C if the radial line through the point intersects C, that is, if some scalar multiple (tx, ty, tz) of (x, y, z) satisfies the equations

$$f(tx, ty) = 0 \text{ and } tz = 1.$$

Solving the equation on the right for t we get $t = 1/z$. Substitute this result into the equation on the left:

$$f\left(\frac{x}{z}, \frac{y}{z}\right) = 0. \tag{4.4}$$

This equation gives the condition for a point (x, y, z) to lie on the projectivized curve if $z \neq 0$.

The projectivized curve may also contain points at infinity on the horizon of the $z = 1$ plane. Points at infinity are radial lines containing nonzero points with vanishing z coordinate. To find the points at infinity, multiply Equation 4.4 through by z^d, where $d = $ degree of f, to clear the fractions. (Multiplying by z^d does not affect the solutions of Equation 4.4 at points where $z \neq 0$). The result,

$$F(x, y, z) = z^d f\left(\frac{x}{z}, \frac{y}{z}\right),$$

is the homogenization of f.

The above process needs to be modified slightly if the Euclidean curve does not lie in the $z = 1$ plane, but the basic approach is the same.

Example 4.7.5 Use the technique of Examples 4.5.1 and 4.5.2 to find homogeneous equations for the projectivizations of the parabolas

$$Q_1 = \left\{ Y = 1 + \frac{X^2}{4}, Z = -1 \right\} \text{ and}$$

$$Q_2 = \left\{ Y = \frac{X^2}{4}, Z = -1 \right\}$$

in the $z = -1$ plane. A point (x, y, z) with $z \neq 0$ lies on the cone through Q_1 if and only if

$$ty = 1 + \frac{(tx)^2}{4} \text{ and } tz = -1$$

for some scalar t. Thus $t = -1/z$ and

$$\frac{-y}{z} = 1 + \frac{1}{4} \left(\frac{-x}{z} \right)^2.$$

Multiply through by z^2 to clear the fractions, and replace the small letters x, y, z by capital letters. The result is a homogeneous equation for the projectivized curve Q_1:

$$YZ + Z^2 + \frac{1}{4}X^2 = 0.$$

A similar computation produces a homogeneous equation for the projectivized curve Q_2:

$$YZ + \frac{1}{4}X^2 = 0.$$

The equations for the original Euclidean curves can be recovered from these equations by replacing X by x, Y by y, and Z by -1.

Exercise 4.7.2 Find a homogeneous polynomial equation for the cone with vertex at the origin that passes through the circle where the plane $X + Y + Z = 1$ intersects the sphere $X^2 + Y^2 + Z^2 = 1$.

4.8 Tangents

The projectivization of a line that is tangent to a Euclidean curve is a radial plane that is tangent to a cone. The cone is the projectivization of the curve. If $F(X, Y, Z) = 0$ is a homogeneous equation for the projectivized curve and at least one of the partial derivatives $\partial F/\partial X$, $\partial F/\partial Y$, $\partial F/\partial Z$ is nonzero at (X_0, Y_0, Z_0), then the equation of its tangent plane at (X_0, Y_0, Z_0) is

$$X\frac{\partial F}{\partial X} + Y\frac{\partial F}{\partial Y} + Z\frac{\partial F}{\partial Z} = 0, \tag{4.5}$$

where the partial derivatives are all evaluated at (X_0, Y_0, Z_0). The same equation defines the projective line that is tangent to the corresponding projective curve at $[X_0, Y_0, Z_0]$, since the tangent plane *is* the projective line (see Fig. 4.23 on page 141).

Lines that are asymptotic to the original Euclidean curve become tangents at points at infinity on its projectivization.

If all of the partial derivatives of F are zero at $[X_0, Y_0, Z_0]$ there may be more than one tangent line at this point, and $[X_0, Y_0, Z_0]$ is said to be a *singular point* on the curve (Fig. 4.24).

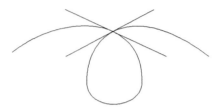

FIGURE 4.24. Tangents at a singular point.

Exercise 4.8.1 a) Consider the hyperbola $x^2/a^2 - y^2/b^2 = 1$ in the $z = 1$ plane in $\mathbf{R}^3$. Find homogeneous equations for the projectivization of the hyperbola and for the projectivizations of its asymptotic lines. Verify directly that the projectivized asymptotic lines are tangent to the projectivized hyperbola where it intersects the line at infinity.

b) Show that projectivization of the parabola $y = x^2$ in the $z = 1$ plane is tangent to the line at infinity.

Exercise 4.8.2 a) Show that the projectivization of the Euclidean curve $x^2 = y(x^2 - y^2)$ intersects the line at infinity at three points: $[X, Y, Z] = [1, 0, 0], [1, 1, 0]$, and $[1, -1, 0]$. (Remember that homogeneous coordinates are defined only up to scalar multiples!)

b) Find the tangents to the projectivized curve at the points where it intersects the line at infinity and use them to find all the asymptotes of the Euclidean curve $x^2 = y(x^2 - y^2)$. Sketch the graph of the curve. (Hint: intersect the projectivized tangents with the $z = 1$ plane).

Exercise 4.8.3 a) Consider a conic given by an equation

$$Ax^2 + Bxy + Cy^2 + Dx + Ey + F = 0$$

in the $z = 1$ plane. Show that the projectivized conic meets the line at infinity at points $[X, Y, 0]$ where

$$\frac{X}{Y} = \frac{-B \pm \sqrt{B^2 - 4AC}}{2A}.$$

b) Use the result from part a) to explain why the conic is

elliptic if $B^2 - 4AC < 0$,
parabolic if $B^2 - 4AC = 0$,
hyperbolic if $B^2 - 4AC > 0$.

4.9   Dual Curves

For every curve $C \subseteq \mathbf{P}^2$ there is a *dual curve* $C^* \subseteq \mathbf{P}^{2*}$. A point on C^* is a line that is tangent to C.

Definition 4.9.1

$$C^* = \{P^* \in \mathbf{P}^{2*} \mid P^* \text{ is tangent to } C\}.^2$$

The Dual of the Dual Curve
The dual C^{**} of the dual curve consists of lines that are tangent to the dual curve. Since a line in $\mathbf{P}^{2*}$ is a point in $\mathbf{P}^2$, C^* is a curve in $\mathbf{P}^2$.

Proposition 4.9.1 $C^{**} = C$.

Sketch of Proof. (See Fig. 4.25). Suppose that C and C^* are smooth curves. For each $P \in C$ and $P^* \in C^*$ set

$T(P) = $ line tangent to C at P,
$T^*(P^*) = $ line tangent to C^* at P^*.

To prove the proposition it is enough to show that

$$P^* = T(P) \text{ if and only if } P = T^*(P^*)$$

for arbitrary P and P^*.
    Assume that $P^* = T(P)$. $T^*(P^*)$ is the limit

$$T^*(P^*) = \lim_{Q^* \to P^*} \overleftrightarrow{Q^* P^*} \tag{4.6}$$

---

[2]If C is singular this definition needs to be modified slightly. In that case C^* also contains the limit of every convergent sequence of lines that are tangent to C.

of secant lines $\overleftrightarrow{Q^*P^*}$ as Q^* goes to P^* on C^*. Since points P^* and Q^* are (infinitesimally close to) lines that are tangent to C we may assume that

$$Q^* = T(Q)$$

for some point $Q \in C$. Regard $T^*(P^*)$ as a point in $\mathbf{P}^2$ and dualize. Equation 4.6 becomes

$$T^*(P^*) = \lim_{Q \to P} T(Q) \cap T(P).$$

But

$$\lim_{Q \to P} T(Q) \cap T(P) = P.$$

Hence $T^*(P^*) = P$. A similar argument shows that $T(P) = P^*$.

This proves Proposition 4.9.1 if C and C^* are smooth. If C or C^* is not smooth then some technicalities involving singular points must be dealt with but in the end everything works out.

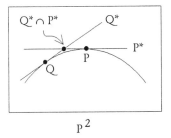
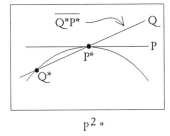

FIGURE 4.25. Dual curves.

The equation of a dual curve

It rarely is necessary to compute the equation of a dual curve. This is fortunate, since the calculation of the equation of a dual curve usually involves some pretty complicated algebra. The following example gives an idea of how it can be done.

Example 4.9.1 Let C be the projectivization of the Euclidean hyperbola

$$\frac{x^2}{4} - \frac{y^2}{9} = 1.$$

To compute the dual of C we start by finding all the lines that are tangent to C. Let L be the line whose equation

$$y = mx + b.$$

L is tangent to the hyperbola if and only if it intersects the hyperbola at only one point and is not parallel to one of the asymptotes. To find the points where L intersects the hyperbola, plug the equation for L into the equation for the hyperbola:

$$\frac{x^2}{4} - \frac{(mx+b)^2}{9} = 1,$$

multiply out:

$$x^2 \left(\frac{1}{4} - \frac{m^2}{9} \right) - x \left(\frac{2mb}{9} \right) - \left(\frac{b^2}{9} + 1 \right) = 0 \qquad (4.7)$$

and solve for x. Equation 4.7 takes the form

$$Px^2 + Qx + R = 0$$

where

$$P = \left(\frac{1}{4} - \frac{m^2}{9} \right), \quad Q = - \left(\frac{2mb}{9} \right), \quad R = - \left(\frac{b^2}{9} + 1 \right). \qquad (4.8)$$

The asymptotic lines of the hyperbola have slope $\pm\sqrt{9/4} = \pm 3/2$. If L is not parallel to an asymptotic line it follows that $P \neq 0$, so one can use the quadratic formula to solve for x:

$$x = \frac{-Q \pm \sqrt{Q^2 - 4PR}}{2P}.$$

There is exactly one solution if and only if

$$Q^2 - 4PR = 0.$$

This is the condition for the line to be tangent to the hyperbola. If you plug in Equations 4.8, it takes the form:

$$\frac{b^2}{9} - \frac{4m^2}{9} + 1 = 0 \qquad (4.9)$$

which is a hyperbola in the m,b plane.

C^* is the projectivization of this hyperbola, because Exercise 4.6.2 shows that the m,b plane is just the set of points $[A, B, C] \in \mathbf{P}^{2*}$ with $B = -1$, so the hyperbola is the image of C^* in the viewplane $\{(A, B, C) \mid B = -1\}$.

Proposition 4.9.2 The dual of a smooth conic in $\mathbf{P}^2$ is a smooth conic in $\mathbf{P}^{2*}$.

Sketch of proof. One can prove this using the same method as in Example 4.9.1, starting with the general equation for a conic (Equation 3.27). We omit the details.

Exercise 4.9.1 Find an equation for the dual of the circle $x^2 + y^2 = 1$.

4.10   Pascal's and Brianchon's Theorems

A *hexagon* is the union of six lines in $\mathbf{P}^2$.

Theorem 4.10.1 (Blaise Pascal [1623-1662]).

> *If a hexagon is inscribed in a smooth conic, the intersections of opposite sides of the hexagon are collinear.*

(In other words if the points A, B, C, A', B', C' lie on a smooth conic then

$$P = \overleftrightarrow{AB'} \cap \overleftrightarrow{A'B}, \quad Q = \overleftrightarrow{AC'} \cap \overleftrightarrow{A'C}, \quad R = \overleftrightarrow{BC'} \cap \overleftrightarrow{B'C}$$

are collinear. See Fig. 4.26 and compare with Pappus' Theorem 4.6.1, page 138).

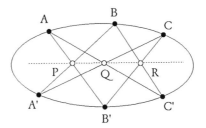

FIGURE 4.26. Pascal's theorem.

Proof. (Adapted from [4, Ch. IV, §8.4, pages 209– 212]. See Fig. 4.27.) Let $X = \overleftrightarrow{AB'} \cap \overleftrightarrow{BC'}$ and $Y = \overleftrightarrow{AC'} \cap \overleftrightarrow{B'C}$. Projecting from B we see that

$$[C', B', A, A'] = [\overleftrightarrow{BC'}, \overleftrightarrow{BB'}, \overleftrightarrow{BA}, \overleftrightarrow{BA'}]$$

by the definition of cross ratio of four points on a conic (Definition 4.4.4). Also

$$[X, B', A, P] = [\overleftrightarrow{BC'}, \overleftrightarrow{BB'}, \overleftrightarrow{BA}, \overleftrightarrow{BA'}]$$

by the definition of the cross ratio of four points on a line (Definition 4.4.3). Hence

$$[C', B', A, A'] = [X, B', A, P]. \tag{4.10}$$

If we project from C the same argument shows that

$$[C', B', A, A'] = [\overleftrightarrow{CC'}, \overleftrightarrow{CB'}, \overleftrightarrow{CA}, \overleftrightarrow{CA'}]$$

and

$$[C', Y, A, Q] = [\overleftrightarrow{CC'}, \overleftrightarrow{CB'}, \overleftrightarrow{CA}, \overleftrightarrow{CA'}],$$

so
$$[C', B', A, A'] = [C', Y, A, Q].$$
Combining this with Equation 4.10, we have
$$[X, B', A, P] = [C', Y, A, Q]. \tag{4.11}$$

Now let $\overleftrightarrow{PQ}$ be the line at infinity, and regard the rest of $\mathbf{P}^2$ as a Euclidean plane. To show that P,Q, and R are collinear it suffices to show that R is a point at infinity, or in other words that $\overleftrightarrow{BC'}$ is parallel to $\overleftrightarrow{B'C}$ in the Euclidean plane. Since P and Q are points at infinity, $\overleftrightarrow{AB'}$ is parallel to $\overleftrightarrow{A'B}$ and $\overleftrightarrow{AC'}$ is parallel to $\overleftrightarrow{A'C}$ in the Euclidean plane. Also since P and Q are points at infinity, Equation 4.11 says that
$$\frac{\overrightarrow{XA}}{\overrightarrow{B'A}} = \frac{\overrightarrow{C'A}}{\overrightarrow{YA}}.$$

Therefore, since each of the triangles $\triangle YAB'$ has one side on $\overleftrightarrow{AB'}$ and another side on $\overleftrightarrow{AC'}$, it follows that these two triangles are similar. Hence
$$\overleftrightarrow{BC'} \text{ is parallel to } \overleftrightarrow{B'C}.$$

In particular $R = \overleftrightarrow{BC'} \cap \overleftrightarrow{B'C}$ is a point at infinity, so it lies on $\overleftrightarrow{PQ}$. This completes the proof.

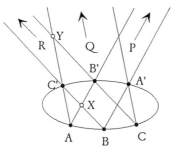

FIGURE 4.27.

Remark. The points on the hexagon need not be arranged exactly as they are in Fig. 4.26 or Fig. 4.27. Pascal's theorem is true no matter how the points are arranged, so long as they all lie on a single smooth conic. The six points need not even be distinct; Pascal's theorem will hold even if two adjacent points on the hexagon are equal, if we take the corresponding edge of the hexagon to be the tangent line at that point.

We get the next theorem for free, by dualizing Pascal's theorem.

Theorem 4.10.2 Brianchon's Theorem. *(Charles Julien Brianchon).*

If a hexgon is circumscribed around a smooth conic, the lines connecting opposite vertices all pass through the same point.

(In other words, given six lines A, B, C, A', B', C' tangent to a smooth conic K the lines

$$P = \overleftrightarrow{(A \cap B')(A' \cap B)},$$

$$Q = \overleftrightarrow{(A \cap C')(A' \cap C)},$$

$$R = \overleftrightarrow{(B \cap C')(B' \cap C)}$$

all intersect at the same point. See Fig. 4.28.)

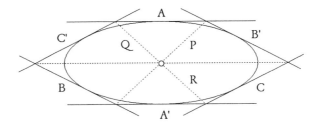

FIGURE 4.28. Brianchon's theorem.

Proof. Dualize Pascal's theorem, using the fact that the dual of a smooth conic is a smooth conic.

Tangents to a Smooth Conic, Revisited.
We constructed tangents to a circle in Exercise 1.9.5 on page 31. The construction in the following claim enables us to construct tangents to any smooth conic.

Claim 4.10.1 *Let K be a smooth conic and O a point outside K. Draw lines L_1, L_2, L_3 through O, each intersecting K in two points. Set*

$$L_1 \cap K = \{A, A'\},$$
$$L_2 \cap K = \{B, B'\},$$
$$L_3 \cap K = \{C, C'\}.$$

Let

$$P = \overleftrightarrow{AB'} \cap \overleftrightarrow{A'B},$$

$$R = \overleftrightarrow{BC'} \cap \overleftrightarrow{B'C},$$

$$S \in K \cap \overleftrightarrow{PR}.$$

Then $\overleftrightarrow{OS}$ is tangent to K at S (Fig. 4.29).

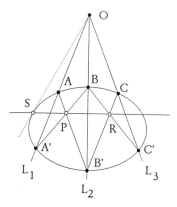

FIGURE 4.29. Tangent to a conic.

Exercise 4.10.1 Prove Claim 4.10.1 in the following two steps:
  a) Set

$$
\begin{aligned}
Q &= \overleftrightarrow{AC'} \cap \overleftrightarrow{A'C}, \\
X &= \overleftrightarrow{AB} \cap \overleftrightarrow{A'B'}, \\
Y &= \overleftrightarrow{AC} \cap \overleftrightarrow{A'C'}, \\
Z &= \overleftrightarrow{BC} \cap \overleftrightarrow{B'C'}.
\end{aligned}
$$

Prove that all six points R, S, T, X, Y, Z are collinear.

(Hint. Apply Pascal's theorem to the hexagon $AB'CA'BC'$. Then apply Desargues' theorem to triangles $\triangle ABC$ and $\triangle A'B'C'$, triangles $\triangle A'BC$ and $\triangle AB'C'$, triangles $\triangle AB'C$ and $\triangle A'BC'$, and triangles $\triangle ABC'$ and $\triangle A'B'C$.)

  b) By part a) $\overleftrightarrow{PR} = \overleftrightarrow{PX}$, so

$$
S = K \cap \overleftrightarrow{PX}.
$$

Observe that P and X depend only on L_1 and L_2. Therefore one can rotate the line L_3 around the point O without affecting S. Rotate L_3 until it becomes tangent, that is, until $C = C'$, and apply part a).

5
Special Relativity

5.1 Spacetime

"The whole of science is nothing more than a refinement of everyday thinking." – Albert Einstein.

"Time is nature's way of preventing everything from happening all at once."

The subject of this chapter is Einstein's special relativity theory and what it says about the geometry of flat spacetime. This is not so difficult or abstruse as it sounds; it involves little beyond high school mathematics. A *spacetime* is simply the mathematical version of a universe that, like our own physical universe, has dimensions both of space and of time. A *flat* spacetime is a spacetime with no gravity, since gravitation tends to "bend" a spacetime. Flat spacetimes are the simplest kind of spacetimes; they stand in the same relation to curved spacetimes as a flat Euclidean plane does to a curved surface.

What makes a flat spacetime different from a Euclidean space is, of course, the existence of a time dimension. In the two-dimensional case there is just one space dimension and one time dimension; we will focus on this case because it is the simplest and yet it illustrates the key elements of the theory.

We will need to discuss lengths of vectors and angles between vectors in a two-dimensional spacetime, and to have an analogue of the Pythagorean theorem. You should not be too surprised when these turn out to behave differently than lengths and angles in Euclidean space. After all, what does it mean to speak of the "length of" or the "angle between" vectors that are pointing forward in time?

Events and Worldlines

In mechanics one studies objects moving through space over intervals of time. Motion along a line is represented by the graph of a function $x = x(t)$ in the t, x plane, a two- dimensional spacetime where t represents time and $x(t)$ represents the position of the object at time t. (Physics texts usually put time on the vertical axis and space on the horizontal axis). The local slope dx/dt of the graph is the velocity of the particle (Fig 5.1).

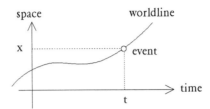

FIGURE 5.1. A two-dimensional spacetime.

An *event* is a point in a spacetime. A *worldline* is a curve $x = x(t)$ that shows the position of a particle as a function of time. The event at time t in the life of the particle occurs at the point (t, x) in the spacetime. The t, x plane has only one space dimension, so worldlines in the t,x plane represent the motions of particles along a line. Motions of particles in higher dimensional spaces require higher dimensional spacetimes. The worldline of a particle moving in the x, y plane is the graph of a function $(x, y) = (x(t), y(t))$ in a three dimensional t,x,y spacetime; for each t the event at the point $(t, x(t), y(t))$ is the event in the life of the particle at time t. The worldline of a particle moving in three dimensional space exists in a spacetime with four independent dimensions t,x,y,z.

Nature does not equip her spacetimes with a set of coordinate axes. In real life coordinates like t and x must be artificially defined and measured by observers who live within the spacetime. Different observers may set up different coordinate systems and disagree on the locations and times of events, so it is important to find a way to relate these different measurements to each other. The search for a solution to this problem leads to the Special Theory of Relativity.

Imagine two observers, O and $\widetilde{O}$, living in a two dimensional spacetime. Observer O assigns coordinates (t, x) to each event by measuring

$t = $ the time (according to O) when the event occurred, and
$x = $ the distance (according to O) between himself and the event
     at the moment when it occurred.

$\widetilde{O}$ also assigns coordinates, which he calls $(\tilde{t}, \tilde{x})$, to each event:

$\tilde{t} =$ the time (according to $\widetilde{O}$) when the event occurred, and
$\tilde{x} =$ the distance (according to $\widetilde{O}$) between himself and the event
at the moment when it occurred.

A Fundamental Problem: What is the relation between the coordinates (t, x) and $(\tilde{t}, \tilde{x})$ of two different observers in the same spacetime?

A basic difficulty arises from the fact that the observers are likely to be moving. Since their state of motion affects their measurements, each observer needs to be able to measure his own motion in order to correct for its effects. But without a natural system of coordinates to refer to, it is physically impossible for the observers to measure, in any absolute sense, exactly where they are or how they are moving, or even whether or not they are moving at all!

Imagine that both observers are coasting along in deep space. If an observer turns on his rocket engine he will be pressed back in his seat. He knows that he is moving because he feels the acceleration. If he is not accelerating he will feel no motion even though he may be moving with great speed. Though he can measure his position and velocity *relative* to other objects in the universe he cannot determine his *absolute* position or velocity since the absolute positions and velocities of these other objects are also unknown.

In the end one is forced to conclude that position and velocity are physically meaningful only in relation to other objects. "Absolute position" and "absolute velocity" are meaningless abstractions. As long as an observer is not accelerating then as far as he can tell or we can tell (if we are moving with him) he might as well be standing still.

An *inertial observer* is an observer who is not accelerating. His or her coordinates are *inertial coordinates*. The basic premise of relativity – that nature does not come equipped with a special set of coordinates and all motion is relative – means that the universe looks the same to all inertial observers: if two inertial observers do the same experiment then they will get the same results. This is:

The Principle of Relativity. The laws of physics are the same in any inertial coordinate system.

5.2 Galilean Transformations

From now on O and $\widetilde{O}$ will be inertial observers moving with a constant relative velocity v so that each observer sees the other moving away from him with velocity v. Figure 5.2 shows what spacetime looks like to O in his own system of coordinates.

Since x measures the distance from O to an event, points where $x = 0$ are events on O's own worldline. The event $(0, 0)$ is the event in the life of

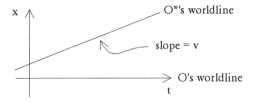

FIGURE 5.2. Spacetime in O's coordinates.

O at time $t = 0$, $(1, 0)$ is the event in the life of O at time $t = 1$, $(2, 0)$ is the event in the life of O at time $t = 2$, etc. Taken together, the points on O's worldline form the t axis.

The x axis is the set of all events with $t = 0$. It is a snapshot of the universe at time $t = 0$. Vertical lines $t = 1$, $t = 2$, $t = 3,...$ are snapshots of the universe at times $t = 1$, $t = 2$, $t = 3,...$

$\tilde{O}$ is traveling with constant velocity v away from O, so his x coordinate at time t is

$$x = vt + x_0 \tag{5.1}$$

where x_0 is a constant. The line $x = vt + x_0$ is $\tilde{O}$'s worldline expressed in O's coordinates; its slope v is the velocity of $\tilde{O}$ relative to O.

Before Einstein, physicists assumed that O and $\tilde{O}$ would get the same result whenever they measured the interval of time between two events. It follows that the time on O's clock can differ from the time on $\tilde{O}$'s clock only by a constant, t_0:

$$\tilde{t} = t + t_0. \tag{5.2}$$

One cannot expect O and $\tilde{O}$ to agree on the positions of events that occur at different times because each observer measures position on his own ruler, which moves with him relative to the other observer. For the same reason, a person standing on the side of a road would not measure the length of a moving car by comparing the position of its rear bumper at one time with the position of its front bumper one minute later.

Nevertheless it was assumed that the observers would agree on the distances between *simultaneous* events. To measure the length of the moving car, you compare the positions of its front and rear bumpers *at the same time.*

Let e be an event with $\tilde{O}$-coordinates $(\tilde{t}, \tilde{x})$. e is $\tilde{x}$ units of distance away from $\tilde{O}$ when $\tilde{O}$'s clock reads $\tilde{t}$. At that moment, O's clock reads $t = \tilde{t} - t_0$. At time t the distance from O to $\tilde{O}$ is $x(t) = vt + x_0$ and the distance from $\tilde{O}$ to e is $\tilde{x}$. Both observers agree on these measurements since they were made simultaneously, so we can combine them to find the distance at that moment from O to e:

$$
\begin{aligned}
x &= (\text{distance from } \tilde{O} \text{ to } e) + (\text{distance from } O \text{ to } \tilde{O}) \\
&= \tilde{x} + vt + x_0.
\end{aligned}
$$

Collecting Equations 5.2 and 5.3 together, we have:

Proposition 5.2.1 The Galilean Transformations.[1]

Let O and $\widetilde{O}$ be a pair of inertial observers moving with relative velocity v, and let (t, x) and $(\tilde{t}, \tilde{x})$ be their respective inertial coordinate systems. If O and $\widetilde{O}$ agree on the time interval between any two events and on the distances between simultaneous events, then

$$\tilde{t} = t + t_0 \text{ and}$$
$$\tilde{x} = x - vt - x_0$$

where t_0 and x_0 are constants.

Figure 5.3 compares the two coordinate systems. The vertical lines are $t = $ constant lines (respectively, $\tilde{t} = $ constant lines) and the horizontal lines are $x = $ constant (respectively $\tilde{x} = $ constant) lines.

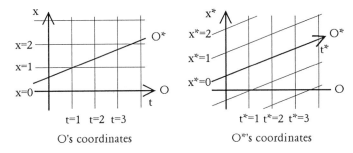

O's coordinates   O*'s coordinates

FIGURE 5.3. Galilean Transformations.

Galilean Spacetimes

Definition 5.2.1 Galilean Spacetime. A *Galilean spacetime* is a spacetime in which the coordinates of inertial observers are related by Galilean transformations. *Galilean observers* are inertial observers in a Galilean spacetime.

Galilean observers agree on the elapsed time between events and on the distance between simultaneous events, and they have a particularly simple formula for the addition of velocities.

Proposition 5.2.2 Addition of Velocities in Galilean Spacetimes.

Let O and $\widetilde{O}$ be Galilean observers and A an object in a Galilean spacetime. If $\widetilde{O}$ is traveling with velocity v relative to O, and A is traveling in the same direction with velocity w relative to $\widetilde{O}$, then A is traveling with velocity $v + w$ relative to O.

---

[1]Named after the famous Italian astronomer and physicist Galileo of Galilei (1564-1642).

In other words, if a child is running forward with a speed of 10 miles per hour inside a train which is traveling at 50 miles per hour then the child's total speed relative to the ground is 60 miles per hour.

Proof. Since A is traveling with constant velocity w relative to $\widetilde{O}$, A's worldline satisfies an equation of the form

$$\tilde{x} = w\tilde{t} + \tilde{x}_0$$

in $\widetilde{O}$'s coordinates, where $\tilde{x}_0$ is a constant. Substitute the Galilean transformations into this equation and get

$$x - vt - x_0 = w(t + t_0) + \tilde{x}_0$$

in $O's$ coordinates. Solve for x to find the position of A at time t according to O:

$$\begin{aligned} x &= (v + w)t + (x_0 + wt_0 + \tilde{x}_0) \\ &= (v + w)t + (\text{constant}). \end{aligned}$$

Hence A's velocity is $v + w$ according to O.

5.3   The Failure of the Galilean Transformations

It turns out that we do not live in a Galilean spacetime. Observers in our universe do not agree on the time interval between events, they do not agree on distances between events that are simultaneous according to some observer, and velocities do not add in the simple way described in Proposition 5.2.2. It took a long time for scientists to realize this because the assumption that our universe is Galilean produces only negligible errors at low relative speeds. But at speeds close to the speed of light the errors are too large to ignore.

By the end of the nineteenth century scientists had measured the speed of light to great accuracy under many different conditions. Experiments[2] show that all observers get the same result, about 186,282 miles per second, when they measure the speed of light in a vacuum regardless of the velocity of the observer or the velocity of the light source.

Suppose a jet fighter is hurtling along at a speed of one mile per second. The pilot measures the speed of light traveling forward in his cockpit to be 186,282 miles per second. According to Galilean velocity addition, a person standing on the ground should see the same light flashing by at a speed of $186,282 + 1 = 186,243$ miles per second. But he does not! In fact, he gets exactly the same figure as the pilot on the plane. The velocity addition formula simply does not work, at least when it is applied to light. Since

---

[2]For instance the famous Michelson-Morely experiments (1881-1887). See [11].

the velocity addition formula is derived from the assumption that different observers could agree on distances between simultaneous events and on time, this assumption must also be false.

5.4   Lorentz Transformations

In this section we enter the strange world of relativistic physics by replacing the Galilean transformations with transformations that are consistent with the fact that the speed of light is the same for all inertial observers.

Simultaneity and the Relativity of Time.

Imagine a pair of flashing lights mounted at points A and B a certain distance, say one mile, apart. Place a detector at a point halfway between the two lights and measure whether or not the light from A arrives at the detector at the same time as the light from B. Both signals travel the same distance at the same velocity, so both take the same amount of time to reach the detector. Therefore we know that A flashed first if the signal from A arrives at the detector before the signal from B, B flashed first if the signal from A arrives later than the signal from B, or A and B flashed simultaneously if the two signals arrive together (Fig. 5.4).

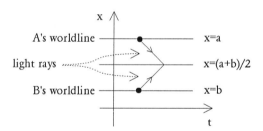

FIGURE 5.4. Simultaneous events.

Definition 5.4.1 Simultaneity of Distant Events.

Let O be an observer with inertial coordinate system (t, x). Two events occur *simultaneously* according to O if and only if two light signals, one originating at each event, would arrive together at the point exactly halfway between the events.

The exact location of the point halfway between the events is determined by O, using his own ruler.

Comments on the definition.

1) The events in Definition 5.4.1 are not actually required to send out any light signals, but whatever means are used to assign times to the events must give the same results as would have been obtained if light signals had been sent.

2) The measurement of simultaneity depends on the observer. It is entirely possible that another observer, carrying out the same measurement, will get different results. Time, as well as position, may be different for different observers.

Let O and $\tilde{O}$ be inertial observers with inertial coordinate systems (t, x) and $(\tilde{t}, \tilde{x})$, respectively, moving apart with relative velocity v. We seek functions f and g which are consistent with the principle of relativity, the definition of simultaneity, and the fact that the speed of light is constant, such that

$$\tilde{t} = f(t, x) \text{ and } \tilde{x} = g(t, x).$$

Assume that f and g have a form that is similar to the Galilean transformations, that is, that there are *constants* p, q, r, s, x_0, and t_0 such that

$$\begin{aligned} f(t, x) &= pt + qx + t_0 \quad \text{and} \\ g(t, x) &= rt + sx + x_0. \end{aligned} \tag{5.3}$$

Besides the fact that this assumption simplifies our task considerably, there are other reasons for asserting that it should be so. It is equivalent to the statement that the partial derivatives $\partial\tilde{t}/\partial t$, $\partial\tilde{t}/\partial x$, $\partial\tilde{x}/\partial t$, and $\partial\tilde{x}/\partial x$ are all constant. If the partials were not constant then we might expect to find a point (t, x) where the partial derivatives were especially nice, say where $\partial\tilde{t}/\partial t$ has a minimum or something of that sort. Such a point would be a specially marked point in spacetime, and we could base a special coordinate system upon it. But the principle of relativity says that there are no specially marked points or other natural features in the universe on which to base a special "natural" system of coordinates, [3] so we conclude that such a point does not exist. Hence the partial derivatives should be constant.

In any case let us agree to accept Equations 5.3. Then

$$\begin{aligned} \tilde{t} &= pt + qx + t_0 \quad \text{and} \\ \tilde{x} &= rt + sx + x_0. \end{aligned} \tag{5.4}$$

Our job is to find the constants p, q, r, s, t_0, and x_0.

To simplify the calculations let us suppose that O and $\tilde{O}$ have some common event in their lives, that is, that their worldlines intersect at some event E. They might as well agree to synchronize their clocks at this event, so we may assume that both observers assign to E the time

$$t(E) = \tilde{t}(E) = 0.$$

---

[3] The special theory of relativity assumes an idealized universe with no gravity and hence no significant masses. The general theory of relativity allows gravity, which bends the surrounding spacetime and destroys its uniformity. In the general theory the partial derivatives $\partial\tilde{x}/\partial x$ etc. are not constant.

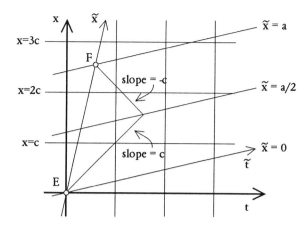

FIGURE 5.5.

Also, because E lies on both of their worldlines, they both assign it the position

$$x(E) = \tilde{x}(E) = 0.$$

Thus O's and $\widetilde{O}$'s coordinate systems have a common origin at the event E. Substituting the coordinates $(0,0) = (t, x) = (\tilde{t}, \tilde{x})$ of E into Equations 5.3 we deduce that

$$t_0 = x_0 = 0. \tag{5.5}$$

Let us also assume that O and $\widetilde{O}$ agree to orient their x axes so that if a light beam is traveling in the positive x direction according to O, then it is also traveling in the positive $\tilde{x}$ direction according to $\widetilde{O}$.

Figure 5.5 is a picture of spacetime in O's coordinates. The t axis is O's worldline and the line $x = vt$ is $\widetilde{O}$'s worldline. Since $\tilde{x} = 0$ on $\widetilde{O}$'s worldline, we have

$$\tilde{x} = 0 \text{ when } x = vt. \tag{5.6}$$

Plug Equations 5.5 and 5.6 into the second of Equations 5.4 to get

$$0 = rt + svt. \tag{5.7}$$

Hence $r = -sv$. Substitute this and Equation 5.5 back into Equations 5.4 to obtain the system

$$\begin{aligned} \tilde{t} &= pt + qx, \\ \tilde{x} &= sx - svt. \end{aligned} \tag{5.8}$$

The next step is to locate the $\tilde{x}$ axis. (The $\tilde{t}$ axis is $\widetilde{O}$'s worldline.) The $\tilde{x}$ axis contains all the events with $\tilde{x}$ coordinate equal to zero, so to find it we must locate another event (other than the origin E) that has $\tilde{t}$

coordinate equal to zero. Let $F = (0, a)$, in $\tilde{O}$'s coordinates, be such an event. According to $\tilde{O}$, F is simultaneous with E since they both occur at time $\tilde{t} = 0$, so light signals sent by E and F will meet at a point halfway between them. E lies on the worldline $\tilde{x} = 0$, F lies on the worldline $\tilde{x} = a$, so the point halfway inbetween lies on the worldline $\tilde{x} = a/2$. Thus to find F all we need to do is follow a light beam from the origin until it crosses the line $\tilde{x} = a/2$, then follow another light beam back from there until it intersects the line $\tilde{x} = a$. (See Fig. 5.5).

Rewrite the equations for these lines in O's coordinates:

$$
\begin{array}{rcllrcl}
\tilde{x} & = & 0 & \text{becomes} & sx - svt & = & 0, \\
\tilde{x} & = & a/2 & \text{becomes} & sx - svt & = & a/2, \\
\tilde{x} & = & a & \text{becomes} & sx - svt & = & a
\end{array}
\tag{5.9}
$$

by Formulas 5.8.

Let

$$
\begin{aligned}
c & = & \text{velocity of light} \\
& \approx & 186,000 \frac{\text{mi}}{\text{sec}}.
\end{aligned}
$$

A light signal traveling from the origin towards $\tilde{x} = a/2$ follows the line $x = ct$, and meets the halfway point $\tilde{x} = a/2$ at the event[4]

$$
(t, x) = \left(\frac{a}{2s(c - v)}, \frac{ac}{2s(c - v)} \right).
$$

(To see this, solve the equations $x = ct$ simultaneously with the second Equation in 5.9.)

A light beam traveling in the opposite direction through this event moves along a line with slope $-c$; it satisfies the equation

$$
x - \frac{a}{2s(c - v)} = -c \left(t - \frac{a}{2s(c - v)} \right)
$$

and meets the line $\tilde{x} = a$ at the event

$$
(t, x) = \left(\frac{av}{s(c^2 - v^2)}, \frac{ac^2}{s(c^2 - v^2)} \right)
$$

by the third of Equations 5.9.

This event is the event F. Since the line connecting F to the origin is the $\tilde{x}$ axis, it follows that the $\tilde{x}$ axis satisfies the equation

$$
\tilde{x} \text{ axis: } x = \frac{c^2}{v} t.
\tag{5.10}
$$

---

[4]Here, and in the rest of this argument, we assume that a is positive.

Returning to the calculation of the constants in Equations 5.4 we have

$$\tilde{t} = 0 \quad \begin{array}{ll} \text{if and only if} & x = -pt/q \\ \text{if and only if} & x = c^2t/v. \end{array}$$

by Formulas 5.8 and 5.10. Hence $-p/q = c^2/v$, so $q = -pv/c^2$, and Equations 5.8 become:

$$\tilde{t} = pt - \frac{pv}{c^2}x \tag{5.11}$$
$$\tilde{x} = sx - svt.$$

Now

$$\tilde{x} = c\tilde{t} \text{ if and only if } x = ct \tag{5.12}$$

because the speed of light equals c for both observers.[5] Substituting in Equations 5.11, Formula 5.12 becomes:

$$sx - svt = cpt - \frac{cpv}{c^2}x \text{ if and only if } x = ct. \tag{5.13}$$

Solve the left hand side for x:

$$x = \frac{cp + sv}{s + \frac{pv}{c}}t.$$

Now comparing the left and right sides of the statement in Equation 5.13, we have

$$\frac{cp + sv}{s + \frac{pv}{c}} = c.$$

Multiply through by $s + pv/c$ to clear the fractions, then simplify to get:

$$p(c - v) = s(c - v).$$

$v \neq c$ since no physical observer can travel at the speed of light. Hence

$$p = s,$$

and Equations 5.11 become

$$\tilde{t} = pt - \frac{pv}{c^2}x \text{ and} \tag{5.14}$$
$$\tilde{x} = px - pvt.$$

It remains only to calculate p.

Imagine an experiment where each observer holds up a ruler that is one foot long for the other observer to measure as he passes by. Both of the

---

[5] And because we have assumed that their x axes are oriented the same way. If their x axes were oriented in opposite ways then $\tilde{x} = -c\tilde{t}$.

observers perform the same experiment: measure the length of a one-foot long ruler that is moving with velocity v. *By the principle of relativity both observers should get the same result.*

To measure the length of the moving ruler each observer records the positions of its endpoints at one particular time, then finds the difference between them. Since the ruler is moving it is important that he record the location of the endpoints at *one* particular time, for if he recorded where the endpoints were at different times the difference in the positions would reflect how far the ruler had moved in the interim in addition to its length.

Suppose that O measures the length of $\widetilde{O}$'s ruler at time $t = 0$. The endpoints of $\widetilde{O}$'s ruler are at $\tilde{x} = 0$ and $\tilde{x} = 1$. In O's coordinates these equations become

$$px - pvt = 0 \text{ and } px - pvt = 1$$

(by Equations 5.14). Substituting in $t = 0$ we find that the endpoints of $\widetilde{O}$'s ruler are located at

$$x = 0 \text{ and } x = \frac{1}{p}$$

when $t = 0$. Thus according to O

$$\text{the length of } \widetilde{O}\text{'s ruler is } \frac{1}{p}. \tag{5.15}$$

Next we calculate the length of O's ruler when it is measured by $\widetilde{O}$ at time $\tilde{t} = 0$. The endpoints of O's ruler are at $x = 0$ and $x = 1$. First substitute $\tilde{t} = 0$ into the first of Equations 5.14 and get $0 = pt - (pv/c^2)x$, so

$$t = \frac{v}{c^2}x.$$

Then plug this result back into the second of Equations 5.14 and get

$$\tilde{x} = p\left(1 - \frac{v^2}{c^2}\right)x.$$

Thus when $x = 0$ we have $\tilde{x} = 0$ and when $x = 1$ we have $\tilde{x} = p(1 - v^2/c^2)$, so

$$\text{the length of } O\text{'s ruler is } p\left(1 - \frac{v^2}{c^2}\right). \tag{5.16}$$

Since both observers performed the same experiment the principle of relativity says that their results (Equations 5.15 and 5.16) must be the same. Hence

$$\frac{1}{p} = p\left(1 - \frac{v^2}{c^2}\right).$$

It follows that

$$p = \frac{1}{\sqrt{1 - \frac{v^2}{c^2}}}. \tag{5.17}$$

Plugging Equation 5.17 into Equations 5.14 we arrive at our final result:

Theorem 5.4.1 Lorentz Transformations.[6]

Let O and $\tilde{O}$ be a pair of inertial observers traveling with relative velocity v. Let (t, x) and $(\tilde{t}, \tilde{x})$ be their respective inertial coordinate systems. Then there are constants t_0, x_0, and $\beta = \pm 1$ such that

$$\tilde{t} = \frac{t - \frac{v}{c^2}x}{\sqrt{1 - \frac{v^2}{c^2}}} + t_0, \text{ and}$$

$$\tilde{x} = \beta\frac{x - vt}{\sqrt{1 - \frac{v^2}{c^2}}} + x_0.$$

The constants t_0 and x_0 in the statement of Theorem 5.4.1 take care of the possibility that O and $\tilde{O}$ might not have the same origin, and β takes care of the fact that they might not orient their x axes the same way. Figure 5.6 shows how O's and $\tilde{O}$'s coordinates compare when they are related by a Lorentz transformation.

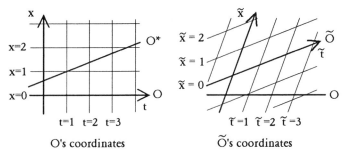

O's coordinates $\tilde{O}$'s coordinates

FIGURE 5.6. Lorentz transformations.

5.5 Relativistic Addition of Velocities

Definition 5.5.1 Minkowski Spacetime.[7]

A *Minkowski Spacetime* is a spacetime in which the coordinates of different inertial observers are related by Lorentz transformations. A *Minkowski observer* is an inertial observer in a Minkowski spacetime.

Proposition 5.5.1 Addition of Velocities in a Minkowski Spacetime. *Let A, B, and C be objects in a Minkowski spacetime. If B is traveling with velocity v relative to A and C is traveling with velocity w relative*

---

[6]Named after their discoverer, Dutch physicist Hendrik Antoon Lorentz (1853-1928).

[7]Hermann Minkowski (1864-1909), German mathematician who reformulated Einstein's Special Realtivity Theory in geometric language.

to B then C is traveling with velocity

$$\frac{v + w}{1 + \frac{vw}{c^2}}$$

relative to A.

Proof. Over a sufficiently short interval of time we may assume these objects have constant velocity, so we may regard them as Minkowski observers. Let (t, x) and $(\tilde{t}, \tilde{x})$ be the inertial coordinates of A and B, respectively.

Since C is traveling with velocity w relative to B, C's worldline satisfies an equation of the form

$$\tilde{x} = w\tilde{t} + \tilde{x}_0$$

in B's coordinates, where $\tilde{x}_0$ is a constant. Plug in the Lorentz transformations and this becomes

$$\frac{x - vt}{\sqrt{1 - \frac{v^2}{c^2}}} + x_0 = w\left(\frac{t - \frac{v}{c^2}x}{\sqrt{1 - \frac{v^2}{c^2}}} + t_0\right) + x_0.$$

Simplifying, we get an equation for the worldline of C in A's coordinates:

$$x = \left(\frac{v + w}{1 + \frac{vw}{c^2}}\right)t + (\text{constant}).$$

Exercise 5.5.1 a) A train is moving at 60 mph. A child on the train runs forward at a speed of 10 mph relative to the train. How fast is she moving relative to the ground?

b) Rework part a) assuming that the train is moving forward at 50% of the speed of light and the child is running forward at 90% of the speed of light relative to the train.

Exercise 5.5.2 The 'Light Barrier'.

a) Let A, B, C, v and w be as in proposition 5.5.1. Show that if $-c < v, w < c$ then the relative velocity of C and A is less than the speed of light.

b) Show that it is impossible in a Minkowski spacetime for an object that is initially travelling slower than the speed of light to accelerate continuously, in a finite amount of its own time, to a speed greater than or equal to the speed of light. (Hint: if $\widehat{O}$'s acceleration, measured by himself, is continuous, then it is bounded over any finite interval $\tilde{t}_0 \leq \tilde{t} \leq \tilde{t}_1$ of his time. Thus over any sufficiently small period $\Delta\tilde{t}$ of time the change $\Delta\tilde{v}$ in his velocity, measured by himself, is less than the speed of light. Now use part a of this exercise, with $w = \Delta\tilde{v}$.)

5.6   Lorentz-FitzGerald Contractions[8]

Moving Clocks Run Slow

One of the strange effects predicted by special relativity is that if two observers are in motion relative to one another then each perceives the other's clock to be moving slower than his own.

Example 5.6.1 The Twin Paradox.

On their twenty-first birthday Paula leaves her twin brother Peter and embarks on a trip at the terrific speed of $(24/25)c$. After traveling for seven years she turns back and returns to Peter at the same speed. Paula is $21 + 2 \times 7 = 35$ years old when she rejoins her twin brother. How old is Peter?

Solution. (See Fig. 5.7). We shall assume that Peter and Paula are inertial observers throughout the first leg of Paula's trip. Let (t, x) be Peter's coordinate system and let $(\tilde{t}, \tilde{x})$ be Paula's. Let $(0,0) = (t,x) = (\tilde{t}, \tilde{x})$ at the event of Paula's departure, so that $0 = t_0 = x_0$ in the Lorentz transformations.

The first leg of Paula's trip ends at $(\tilde{t}, \tilde{x}) = (7, 0)$. Plug $(\tilde{t}, \tilde{x}) = (7, 0)$ and $v = (24/25)c$ into the Lorentz transformations:

$$7 = \frac{t - \frac{24x}{25c}}{\sqrt{1 - \left(\frac{24}{25}\right)^2}}$$

$$0 = \frac{x - \frac{24}{25}ct}{\sqrt{1 - \left(\frac{24}{25}\right)^2}},$$

so

$$(t, x) = (25, 24c)$$

at the end of the first leg of Paula's trip. Therefore the first leg of Paula's trip takes twenty five (of Peter's) years. The second leg takes an equal amount of time so Peter is $21 + 2 \times 25 = 71$ years old when Paula returns.

Although it seems strange that twins could have different ages, there is nothing contradictory about the "twin paradox". Experiments show that the world really works this way. One might ask, why can't the same argument be used to prove that Paula is older than Peter, since she sees him traveling away from her? The answer is that the realtion between Peter and Paula is not really symmetrical: Paula experiences a terrific acceleration when she turns around and heads back, but Peter does not.

---

[8]George FitzGerald (1851-1901), Irish physicist who first proposed that objects contract in the direction of their motion.

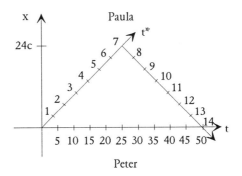

FIGURE 5.7. Twin paradox.

Exercise 5.6.1 Assume that it took Paula a negligible amount of time to reverse her direction of travel and start back.

a) An instant before Paula turned back, how old was she according to Peter and how old was he according to her?

b) An instant after Paula turned back, how old was she according to Peter and how old was he according to her?

c) According to Paula how much did Peter age while Paula was turning around? How much did Paula age according to Peter?

Moving Rulers are Short

Another affect predicted by special relativity is the contraction of space in the direction of motion. The argument leading to Equation 5.17 shows that if a ruler L feet long is moving with velocity v relative to an observer, then the observer will measure its length to be

$$L\sqrt{1 - \frac{v^2}{c^2}}$$

feet long.[9]

Example 5.6.2 The Einstein Train. A train whose length is 200 m. according to an observer on the train passes through a station whose length is 100 m. according to an observer in the station. The speed of the train is $(\sqrt{3}/2)c$. To a man in the station the length of the train is $(200\text{m.})\sqrt{1 - \left(\frac{\sqrt{3}}{2}\right)^2} = 100\text{m}$. So the train has the same length as the station and the two fit together perfectly. But to a man on the train the station is $(100\text{m.})\sqrt{1 - \left(\frac{\sqrt{3}}{2}\right)^2} = 50\text{m}$. long. This is only one quarter of the length of the train, so the station is too short to hold the train.

---

[9]This contraction is only in the direction of motion. Lengths perpendicular to the direction of motion retain their original size.

5.7   Minkowski Geometry

The Dot Product

The *dot product* of two vectors $\overrightarrow{x} = (x_1, \ldots, x_n)$ and $\overrightarrow{y} = (y_1, \ldots, y_n)$ in $\mathbf{R}^n$ is defined by the formula

$$\overrightarrow{x} \cdot \overrightarrow{y} = x_1 y_1 + x_2 y_2 + \cdots + x_n y_n.$$

The dot product computes lengths and angles:

$$|\overrightarrow{x}| = \sqrt{\overrightarrow{x} \cdot \overrightarrow{x}}, \tag{5.18}$$

$$\angle(\overrightarrow{x}, \overrightarrow{y}) = \arccos\left(\frac{\overrightarrow{x} \cdot \overrightarrow{y}}{|\overrightarrow{x}||\overrightarrow{y}|}\right),$$

and it is preserved by Euclidean isometries. In fact it is not hard to prove that a function $f : \mathbf{R}^n \to \mathbf{R}^n$ is a Euclidean isometry if and only if $\overrightarrow{f(A)f(B)} \cdot \overrightarrow{f(C)f(D)} = \overrightarrow{AB} \cdot \overrightarrow{CD}$ for all points $A, B, C, D \in \mathbf{R}^n$. (The proof is left to the reader).

The dot product is so fundamental for the Euclidean geometry of $\mathbf{R}^n$ that it is often said that Euclidean geometry is simply the geometry of the dot product.

The Minkowski Product

Minkowski Geometry is the geometry of the Minkowski product.

Definition 5.7.1 The *Minkowski product* $\overrightarrow{x} * \overrightarrow{y}$ of two vectors $\overrightarrow{x} = (s, x_1, x_2, \ldots, x_n)$ and $\overrightarrow{y} = (t, y_1, y_2, \ldots, y_n)$ in $\mathbf{R}^{n+1}$ is defined by

$$\overrightarrow{x} * \overrightarrow{y} = -c^2 st + x_1 y_1 + x_2 y_2 + \cdots + x_n y_n$$

where c is a constant[10].

A n+1 dimensional *Minkowski space* $\mathbf{M}^n$ is $\mathbf{R}^{n+1}$ together with the Minkowski product. Physically t and $x_1, \ldots, x_n$ are regarded as inertial coordinates. From now on we will concentrate on $\mathbf{M}^2$.

Definition 5.7.2 Generalized Lorentz Transformation. A *generalized Lorentz transformation*[11] on $\mathbf{M}^2$ is a function of the form

$$f(t, x) = \left(\alpha \frac{t - \frac{v}{c^2}x}{\sqrt{1 - \frac{v^2}{c^2}}} + t_0, \ \beta \frac{x - vt}{\sqrt{1 - v^2 c^2}} + x_0\right)$$

---

[10]c = speed of light.
[11]The definition is generalized to include functions like $f(t, x) = (-t, x)$ that reverse the direction of time.

where

$$t_0, x_0, \text{ and } -c < v < c \text{ are constants}$$

and

$$\alpha, \beta = \pm 1.$$

Proposition 5.7.1 Generalized Lorentz Transformations are Minkowski Isometries.

If A, B, C, D are any events in $\mathbf{M}^2$ and f is a generalized Lorentz transformation then

$$\overrightarrow{f(A)f(B)} * \overrightarrow{f(C)f(D)} = \overrightarrow{AB} * \overrightarrow{CD}.$$

Proof. Set

$$
\begin{array}{cccc}
\overrightarrow{AB} & = & (t, x), & \overrightarrow{CD} & = & (t', x'), \\
\overrightarrow{f(A)f(B)} & = & (\tilde{t}, \tilde{x}), & \overrightarrow{f(C)f(D)} & = & (\tilde{t}', \tilde{x}').
\end{array}
$$

If $A = (t_1, x)$ and $B = (t_2, x_2)$ then $\overrightarrow{AB} = B - A$ so

$$(t, x) = (t_2 - t_1, x_2 - x_1).$$

If f is a generalized Lorentz transformation

$$\overrightarrow{f(A)f(B)} = f(B) - f(A) \tag{5.19}$$

$$= \left(\alpha \frac{t - \frac{v}{c^2} x}{\sqrt{1 - \frac{v^2}{c^2}}}, \beta \frac{x - vt}{\sqrt{1 - \frac{v^2}{c^2}}} \right)$$

for some constant $\alpha, \beta = \pm 1$ and v. (The constants t_0 and x_0 in definition 5.7.2 cancel out when you subtract $f(A)$ from $f(B)$). Similarly,

$$\overrightarrow{f(C)f(D)} = \left(\alpha \frac{t' - \frac{v}{c^2} x'}{\sqrt{1 - \frac{v^2}{c^2}}}, \beta \frac{x' - vt'}{\sqrt{1 - \frac{v^2}{c^2}}} \right). \tag{5.20}$$

Plugging in Equations 5.19 and 5.20 and multiplying out, we have

$$-c^2 \tilde{t}\tilde{t}' + \tilde{x}\tilde{x}' = -c^2 tt' + xx', \tag{5.21}$$

which proves the proposition.

(The converse to Proposition 5.7.1 is also true: if f preserves the Minkowski product then f is a generalized Lorentz transformation. We will not use this fact but the interested reader may wish to prove it for her- or himself.)

Corollary 5.7.1 *The Minkowski product has the same form in any inertial coordinate system.*

Proof. This is Equation 5.21.

Whereas there is only one type of nonzero vector in Euclidean geometry, Minkowski geometry has three:

Definition 5.7.3 A nonzero vector $\overrightarrow{A} \in \mathbf{M}^2$ is
  i) *spacelike* if $\overrightarrow{A} * \overrightarrow{A} > 0$,
  ii) *lightlike* if $\overrightarrow{A} * \overrightarrow{A} = 0$,
  iii) *timelike* if $\overrightarrow{A} * \overrightarrow{A} < 0$.

Every timelike vector has a *time orientation* which determines whether it points toward the future or toward the past.

Definition 5.7.4 Time Orientation.
  A timelike vector (t, x) is
  i) *future pointing* if $t > 0$,
  ii) *past pointing* if $t < 0$.

For example the vector $(0, 1)$ is spacelike, $(1, c)$ is lightlike, and $(1, 0)$ is timelike and future pointing. More generally, a vector is timelike if it is parallel to the time axis of some observer, spacelike if it is parallel to the space axis of some observer, and lightlike if it is parallel to the worldline of a light ray (Fig. 5.8).

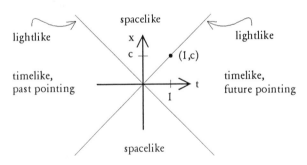

FIGURE 5.8. Time orientation.

Definition 5.7.5 The *worldline of a material particle* in Minkowski space is a parametrized curve whose velocity vectors all are timelike and future pointing.

Material particles have timelike velocity vectors $\overrightarrow{v} = (t, x)$ since their speed $|x/t|$ is less than $|c|$.

Exercise 5.7.1 Show that if $\overrightarrow{A_1}, \overrightarrow{A_2}, \ldots, \overrightarrow{A_n} \in \mathbf{M}^2$ are timelike future pointing vectors then $\overrightarrow{A_1} + \overrightarrow{A_2} + \cdots + \overrightarrow{A_n}$ also is a timelike future pointing vector.

Exercise 5.7.2 a) Prove that if $\overrightarrow{A}$ is a timelike future pointing vector then the $\tilde{t}$ coordinate of A is positive in any inertial coordinate system. (Hint: use Corollary 5.7.1.) State and prove a similar result for past pointing vectors.

Conclude that if $\overrightarrow{ee'}$ is a timelike future pointing vector then *every* observer says that the event e occured before the event e'.

b) Show that if $\overrightarrow{ee'}$ is spacelike then some observers say that e occurs before e', others say the two events occur at the same time, and still other observers say that e' occurs before e.

Exercise 5.7.3 Two timelike vectors $\overrightarrow{A}$ and $\overrightarrow{B}$ have the same *time orientation* if both of them are future pointing, or both are past pointing. Let $\overrightarrow{A}$ and $\overrightarrow{B}$ be timelike vectors. Show that $\overrightarrow{A}$ and $\overrightarrow{B}$ have the same time orientation if and only if both vectors are timelike and $\overrightarrow{A} * \overrightarrow{B} < 0$.

Length

Definition 5.7.6 Length. The *length* of a vector $\overrightarrow{A}$ in Minkowski space is $|\overrightarrow{A}| = \sqrt{|\overrightarrow{A} * \overrightarrow{A}|}$.

(This is a funny kind of "length" since lightlike vectors have zero length even though they are not zero).

The length of a spacelike vector measures distance, but the length of a timelike vector measures *time*.

Proposition 5.7.2 *Let* $e, e' \in \mathbf{M}^2$.

a) If $\overrightarrow{ee'}$ is parallel to the time axis of some observer then $(1/c)|\overrightarrow{ee'}|$ is the time between the two events for that observer.

b) If $\overrightarrow{ee'}$ is parallel to the space axis of some observer then $|\overrightarrow{ee'}|$ is the distance between the two events for that observer.

Proof. If $\overrightarrow{ee'}$ is parallel to the $\tilde{t}$ axis then $\overrightarrow{ee'} = (\tilde{t}, 0)$ for some $\tilde{t}$, and its length is $\sqrt{|-c^2\tilde{t}^2 + 0^2|} = c|\tilde{t}|$. If $\overrightarrow{ee'}$ is parallel to the $\tilde{x}$ axis then $\overrightarrow{ee'} = (0, \tilde{x})$ for some $\tilde{x}$ and its length is $\sqrt{|-c^2 0^2 + \tilde{x}^2|} = |x|$.

Definition 5.7.7 $\overrightarrow{A}$ is *perpendicular* to $\overrightarrow{B}$ in $\mathbf{M}^2$ if $\overrightarrow{A} * \overrightarrow{B} = 0$ (Fig. 5.9).

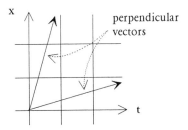

FIGURE 5.9. Minkowski-perpendicular.

Exercise 5.7.4 The Minkowski "Pythagorean Theorem".

Let $\overrightarrow{A}, \overrightarrow{B}, \overrightarrow{C}$ be the sides of a triangle in Minkowski spacetime. Suppose $\overrightarrow{A}$ is timelike, $\overrightarrow{B}$ is spacelike, and $\overrightarrow{A}$ is perpendicular to $\overrightarrow{B}$. Show that

$$-c^2|A|^2 + |B|^2 = |C|^2$$

(see Fig. 5.10).

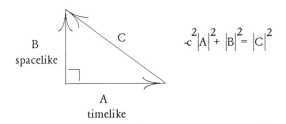

FIGURE 5.10. Minkowski's Pythagorean Theorem.

Exercise 5.7.5 The Minkowski "Triangle Inequality".

Let $\overrightarrow{A}$ and $\overrightarrow{B}$ be future pointing timelike vectors. Prove that

$$|\overrightarrow{A} + \overrightarrow{B}| \geq |\overrightarrow{A}| + |\overrightarrow{B}|$$

with equality if and only if $\overrightarrow{A}$ and $\overrightarrow{B}$ are parallel. What would be different about this result if the triangle were in Euclidean space? What does the triangle inequality have to do with the twin paradox (Example 5.6.1)?

(Hint: see Fig. 5.11. Let $(a, b) = \overrightarrow{A} + \overrightarrow{B}$ be O's coordinates. By exercise 5.7.1 $-c < b/a < c$, so there exists an observer $\widetilde{O}$ who is traveling with velocity $v = b/a$ relative to O. Check that $\overrightarrow{A} + \overrightarrow{B}$ is parallel to $\widetilde{O}$'s time axis. If you write the vectors out in $\widetilde{O}$'s coordinates the exercise will be easy.)

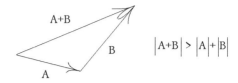

(all timelike, future pointing)

FIGURE 5.11. Minkowski's Triangle Inequality.

5.8   The Slowest Path is a Line

The Length of a Worldline

Let $\alpha(s) = (t(s), x(s))$, $a \leq s \leq b$, parametrize the worldline of a (not necessarily inertial) observer. As s ranges over an infinitesimally small interval Δs, the event $\alpha(s)$ is displaced by an amount

$$\Delta\alpha(s) \approx \alpha'(s)\Delta s.$$

Over a small enough interval we can treat the observer as an inertial observer since his velocity changes by a negligible amount. Thus by part a) of Proposition 5.7.2 the amount of time that elapses on the observer's clock as s varies over the interval Δs is

$$\begin{aligned}\Delta t &\approx \frac{1}{c}|\Delta\alpha| \\ &\approx \frac{1}{c}|\alpha'(s)|\Delta s\end{aligned}$$

with equality in the limit as $\Delta s \to 0$. Hence the total time that elapses on the observer's clock between the event $\alpha(a)$ where $s = a$ and the event $\alpha(b)$ where $s = b$ is

$$\begin{aligned}T &= \frac{1}{c}\int_a^b |\alpha'(s)|ds \qquad (5.22) \\ &= \frac{1}{c}(\text{length of the worldline}).\end{aligned}$$

Time Maximization

The worldline of an inertial observer is a straight line in Minkowski space. In Euclidean geometry straight lines minimize length, but in Minkowski geometry straight lines pointing in a timelike direction *maximize time*.

Proposition 5.8.1 The Generalized Twin Paradox. *An inertial observer takes longer to get from one event to another than any other observer.*

Proof. Let O and $\widetilde{O}$ be observers whose worldlines contain the events e and e'. Assume that O is an inertial observer and $\widetilde{O}$ is not.

In O's coordinates let

$$e = (t_0, 0) \text{ and } e' = (t_1, 0)$$

with $t_0 < t_1$. In $\tilde{O}$'s coordinates, let

$$e = (\tilde{t}_0, 0) \text{ and } e' = (\tilde{t}', 0).$$

We must prove that

$$|t_1 - t_0| > |\tilde{t}_1 - \tilde{t}_0|.$$

To compute $|\tilde{t}_1 - \tilde{t}_0|$, let $\alpha(s) = (t(s), x(s))$, $a \le s \le b$, be a parametriza-tion of $\tilde{O}$'s worldline between e and e'. Then

$$t(a) = t_0 \text{ and } t(b) = t_1,$$

and, since the velocity vector $(t'(s), x'(s))$ of the worldline of a material particle is a future pointing vector (Definition 5.7.5), we also have

$$t'(s) > 0 \text{ for all } s.$$

By the length formula (Equation 5.22),

$$
\begin{aligned}
|\tilde{t}_1 - \tilde{t}_0| &= \frac{1}{c} \int_a^b |\alpha'(s)| ds \\
&= \frac{1}{c} \int_a^b |(t'(s), x'(s))| ds.
\end{aligned}
$$

Because $\tilde{O}$ cannot travel faster than light,

$$-c < \frac{dx}{dt} = \frac{dx/ds}{dt/ds} < c,$$

so

$$c^2 \left(\frac{dt}{ds}\right)^2 > \left(\frac{dx}{ds}\right)^2$$

Hence

$$
\begin{aligned}
|(t'(s), x'(s))| &= \sqrt{|-c^2(t'(s))^2 + (x'(s))^2|} \\
&= \sqrt{c^2(t'(s))^2 - (x'(s))^2} \\
&\le c|t'(s)| \\
&= ct'(s)
\end{aligned}
$$

for every s. It follows that

$$
\begin{aligned}
|\tilde{t}_1 - \tilde{t}_0| &= \frac{1}{c} \int_a^b |(t'(s), x'(s))| ds \\
&\le \frac{1}{c} \int_a^b t'(s) ds \\
&= t(b) - t(a) \\
&= t_1 - t_0
\end{aligned}
$$

with equality on the second line only if $x'(s) = 0$ for all s. Thus O takes longer to travel from e to e' than $\tilde{O}$ does. This completes the proof.

Exercise 5.8.1 Let e and e' be any two events such that $\overrightarrow{ee'}$ is a timelike, future pointing vector. Show that, given any positive number t, there is an observer traveling slower than the speed of light (although not necessarily at a constant speed) who gets from e to e' in less than t units of time, as measured on the observer's own clock.

Thus, traveling slower than the speed of light, you can get from here to Alpha Centauri, 4.3 light years away, in less time than it takes to read this sentence (if you are willing to endure some pretty wild accelerations).

5.9   Hyperbolic Angles and the Velocity Addition Formula

Definition 5.9.1 A *pseudocircle* in Minkowski space is a set of vectors $\overrightarrow{A}$ satisfying an equation of the form

$$|\overrightarrow{A}| = r$$

where r is a constant (fig. 5.12).

If $r \neq 0$ the pseudocircle consists of a pair of hyperbolas

$$-c^2 t^2 + x^2 = \pm r^2.$$

The branch of the hyperbola $-c^2 + x^2 = -r^2$ where $t > 0$ is parametrixed by

$$t = \frac{r}{c} \cosh \phi$$
$$x = r \sinh \phi.$$

A radial vector (t, x) is timelike and future pointing if it extends out to a point on this branch of the hyperbola. The hyperbolic tangent measures the velocity of a particle whose worldline is parallel to the vector (t, x)

$$\tanh \phi = \frac{x}{ct} = \frac{v}{c}. \tag{5.23}$$

Definition 5.9.2 Let $\overrightarrow{A}$ and $\overrightarrow{B}$ be timelike, future pointing vectors. The *hyperbolic angle* ϕ between $\overrightarrow{A}$ and $\overrightarrow{B}$ is defined up to a $\pm$ sign by the formula

$$\cosh \phi = \frac{-\overrightarrow{A} * \overrightarrow{B}}{c^2 |\overrightarrow{A}||\overrightarrow{B}|}.$$

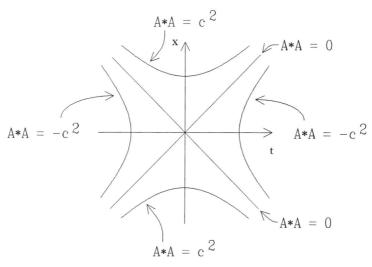

FIGURE 5.12. Pseudocircles.

Lorentz transformations preserve hyperbolic angles because they preserve the lengths and Minkowski products that are used to define them. If $\overrightarrow{A}$ and $\overrightarrow{B}$ are each tangent to the worldline of an observer at some event e then Equation 5.23 says that $\tanh \phi$ is $1/c$ times their relative velocity, because one can always regard the coordinates in Equation 5.23 as being the inertial coordinates of one of the observers.

Given objects A, B, and C let

$\alpha = $ the hyperbolic angle between A's and B's worldlines,
$\beta = $ the hyperbolic angle between B's and C's worldlines,
$\gamma = $ the hyperbolic angle between A's and C's worldlines.

The Lorentz velocity addition formula (Proposition 5.5.1) says that

$$c \tanh \gamma \;=\; \frac{c \tanh \alpha + c \tanh \beta}{1 + \tanh \alpha \tanh \beta}$$
$$=\; c \tanh(\alpha + \beta),$$

where the second equation comes from the addition formula for hyperbolic tangents (see Exercise 5.10.1 in the next section). Thus the velocity addition formula boils down to the statement that

$$\gamma = \alpha + \beta.$$

In this way we recover something very much like the simplicity of the Galilean velocity addition formula on page 158.

5.10   Appendix: Circular and Hyperbolic Functions

If $s = 1$ or -1 the curve

$$x^2 + sy^2 = 1 \tag{5.24}$$

is a circle or a hyperbola in the x,y plane. If $s = 1$ it is a unit circle and

$$x = \cos\phi, y = \sin\phi$$

where ϕ is the angle between the vectors $(1,0)$ and (x,y), measured in radians. The area of the circular sector that is subtended by the vectors $(1,0)$ and (x,y) is given by the formula

$$\frac{\text{area of the sector}}{\text{area of the circle}} = \frac{\phi}{2\pi}.$$

Hence

$$\phi = 2(\text{area of the sector}) \tag{5.25}$$

(see Fig. 5.13).

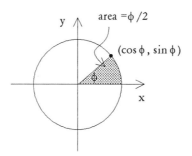

FIGURE 5.13. Circular angle.

If $s = -1$ the curve is a hyperbola. Let (x, y) be a point on the hyperbola with $x > 0$. Define the *hyperbolic angle* ϕ by

$$\phi = 2 \left(\begin{array}{l} \text{area of the sector between the ra-} \\ \text{dial vectors } (0, 1) \text{ and } (x, y) \end{array} \right) \tag{5.26}$$

(see Fig. 5.14), and the hyperbolic cosine, hyperbolic sine and hyperbolic tangent by

$$x = \cosh\phi, \quad y = \sinh\phi, \quad \tanh\phi = \frac{\sinh\phi}{\cosh\phi}.$$

With these definitions the basic identities

$$\cos^2\phi + \sin^2\phi = 1$$
$$\cosh^2\phi - \sinh^2\phi = 1$$

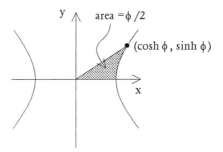

FIGURE 5.14. Hyperbolic angle.

are an immediate consequence of Formula 5.24.

Derivatives of Circular and Hyperbolic Functions

Fix a point (x_0, y_0) on the curve. Let L be the line segment connecting $(0,0)$ to (x_0, y_0) and let M be the horizontal line through (x_0, y_0). The area between the y axis and the hyperbola as y runs from 0 to y_0 is

$$\int_0^{y_0} \sqrt{1 - sy^2}\, dy$$

and the area of the triangle with vertices $(0,0)$, $(0, y_0)$, and (x_0, y_0) is

$$\frac{1}{2} y_0 \sqrt{1 - s y_0{}^2}$$

(Fig. 5.15). The area of the sector between the radial vectors $(0, 1)$ and (x, y) is the difference of these two areas. Combining this with Equation 5.26 we obtain a formula for ϕ as a function of y_0:

$$\phi(y_0) = 2 \int_0^{y_0} \sqrt{1 - sy^2}\, dy - y_0 \sqrt{1 - s y_0{}^2}. \tag{5.27}$$

Proposition 5.10.1

a) $$\frac{d\phi}{dy} = \frac{1}{\sqrt{1 - sy^2}},$$

b) $$\frac{dy}{d\phi} = x,$$

c) $$\frac{dx}{d\phi} = -sy.$$

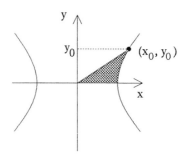

FIGURE 5.15. (See Formula 5.27).

Proofs.
a) Differentiate Equation 5.27 with respect to y_0.

$$\frac{d\phi}{dy_0} = \frac{1}{2}\frac{d(\text{area of sector})}{dy_0}$$

$$= 2\sqrt{1 - sy_0^2} - \sqrt{1 - sy_0^2} + \frac{sy_0^2}{\sqrt{1 - y_0^2}}$$

$$= \frac{1}{\sqrt{1 - sy_0^2}}.$$

b) By a), $dy/d\phi = \sqrt{1 - sy^2} = x$ since $x^2 + sy^2 = 1$.
c) Differentiate the formula $x^2 + sy^2 = 1$ implicitly with respect to ϕ:

$$2x\frac{dx}{d\phi} + 2sy\frac{dy}{d\phi} = 0,$$

so by a),

$$2x\frac{dx}{d\phi} + 2sxy = 0.$$

Then solve for $dx/d\phi$.

Corollary 5.10.1

$$\frac{d\sin\phi}{d\phi} = \cos\phi, \quad \frac{d\cos\phi}{d\phi} = -\sin\phi, \quad \frac{d\sin^{-1}y}{dy} = \frac{1}{\sqrt{1 - y^2}}.$$

Proof. Set $s = 1$ in Proposition 5.10.1. Then $y = \sin\phi$, $x = \cos\phi$, and $\phi = \sin^{-1}y$.

Corollary 5.10.2

$$\frac{d\sinh\phi}{d\phi} = \cosh\phi, \quad \frac{d\cosh\phi}{d\phi} = \sinh\phi, \quad \frac{d\sinh^{-1}y}{dy} = \frac{1}{\sqrt{1 + y^2}}.$$

Proof. Set $s = -1$ in Proposition 5.10.1. Then $y = \sinh\phi$, $x = \cosh\phi$, and $\phi = \sinh^{-1} y$.

Corollary 5.10.3

$$\cosh\phi = \frac{e^\phi + e^{-\phi}}{2}, \quad \sinh\phi = \frac{e^\phi - e^{-\phi}}{2}.$$

Proof. Let $s = -1$, $u = x + y$, and $v = x - y$. By Proposition 5.10.1,

$$\frac{du}{d\phi} = u, \qquad \frac{dv}{d\phi} = -v.$$

It follows that $u = C_1 e^\phi$ and $v = C_2 e^{-\phi}$ for some constants C_1 and C_2. From Fig. 5.14 it is clear that $x = 1$ and $y = 0$ when $\phi = 0$. Hence $u = v = 1$ when $\phi = 0$ so $C_1 = C_2 = 1$. Thus $u = e^\phi$ and $v = e^{-\phi}$.

Exercise 5.10.1 Use the formulas in Corollary 5.10.3 to prove the addition formulas for hyperbolic functions:

a) $\cosh(A + B) = \cosh A \cosh B + \sinh A \sinh B$,

b) $\sinh(A + B) = \sinh A \cosh B + \cosh A \sinh B$,

c) $\tanh(A + B) = \dfrac{\tanh A + \tanh B}{1 + \tanh A \tanh B}$.

References

[1] Arnol'd, V.I., *Mathematical Methods of Classical Mechanics*, 2nd ed., (translated by Vogtmann, K., and Weinstein, A.), Springer-Verlag, 1989.

[2] Artzy, Rafael, *Linear Geometry*, Addison-Wesley, 1974.

[3] Brink, Raymond, *Spherical Trigonometry*, Appleton-Century Inc., N.Y., 1942.

[4] Courant, Richard and Robbins, Herbert, *What is Mathematics*, Oxford University Press, 1978.

[5] Einstein, Albert, *Relativity, the Special and General Theory*, Crown Publishers, N.Y., 1961.

[6] Euclid, *The Elements*, (translated by Sir Thomas Heath), Dover, 1956.

[7] Eves, Howard, *Great Moments in Mathematics (before 1650)*, Mathematical Association of America, 1980.

[8] Feynman, Richard, *QED, The Strange Theory of Light and Matter*, Princeton University Press, 1985.

[9] Francis, George K., *A Topological Picturebook*, Springer-Verlag, 1987.

[10] French, Thomas and Vierck, Charles, *Engineering Drawing*, McGraw-Hill, 1953.

[11] Goldberg, Stanley, *Understanding Relativity*, Birkhäuser Boston, Inc., 1984.

[12] Hartshorne, Robin, *Foundations of Projective Geometry*, W.A. Benjamin, Inc., 1967.

[13] Hilbert, David. and Cohn-Vossen, S. *Geometry and the Imagination*, Chelsea, 1952.

[14] Klein, Felix, *Elementary Mathematics from an Advanced Standpoint; Geometry* (translated by Hedrick, Earle R. and Noble, Charles A.), Dover, 1948.

[15] Kordemsky, B.A., *The Moscow Puzzles*, Charles Scribner's Sons, 1972.

[16] Ogilvy, C. Stanley, *Excursions in Geometry*, Oxford University Press, 1969.

[17] O'Neill, Barrett, *Elementary Differential Geometry*, Academic Press, 1966.

[18] Pedoe, Dan. Geometry and the Liberal Arts. St. Martin's Press. 1979.

[19] Schelin, Charles W., Calculator Function Approximation. *The American Mathematical Monthly* **90** (May 1983), 317–325.

[20] Schumann, Charles H., *Descriptive Geometry*, D. Van Nostrand Co., Inc., 1927.

[21] Thompson, J.E. *Geometry for the Practical Man*, D. Van Nostrand Co., Inc., N.Y., 1934.

[22] Yaglom, I.M.,*Geometric Transformations I*, (translated by Shields, Allen) The New Mathematical Library, Random House, 1962.

Index

Universitext *(continued)*

McCarthy: Introduction to Arithmetical Functions
Meyer: Essential Mathematics for Applied Fields
Meyer-Nieberg: Banach Lattices
Mines/Richman/Ruitenburg: A Course in Constructive Algebra
Moise: Introductory Problem Course in Analysis and Topology
Montesinos: Classical Tessellations and Three Manifolds
Nikulin/Shafarevich: Geometries and Groups
Øksendal: Stochastic Differential Equations
Porter/Woods: Extensions and Absolutes of Hausdorff Spaces
Rees: Notes on Geometry
Reisel: Elementary Theory of Metric Spaces
Rey: Introduction to Robust and Quasi-Robust Statistical Methods
Rickart: Natural Function Algebras
Rotman: Galois Theory
Rybakowski: The Homotopy Index and Partial Differential Equations
Samelson: Notes on Lie Algebras
Schiff: Normal Families of Analytic and Meromorphic Functions
Shapiro: Composition Operators and Classical Function Theory
Smith: Power Series From a Computational Point of View
Smoryński: Logical Number Theory I: An Introduction
Smoryński: Self-Reference and Modal Logic
Stanišić: The Mathematical Theory of Turbulence
Stillwell: Geometry of Surfaces
Stroock: An Introduction to the Theory of Large Deviations
Sunder: An Invitation to von Neumann Algebras
Tondeur: Foliations on Riemannian Manifolds
Verhulst: Nonlinear Differential Equations and Dynamical Systems
Zaanen: Continuity, Integration and Fourier Theory